MANAGING CITIES IN AUSTERITY

URBAN INNOVATION

Urban Innovation stems from the Fiscal Austerity and Urban Innovation Project (FAUI), the most extensive study to date of local government. Made up of 51 international research teams, the Project documents and analyzes the adoption of innovations by local governments, combining a large scale sophisticated research effort with decentralized data collection, interpretation, and policy analysis.

This series includes monographs on a particular theme, such as privatization of urban services, as well as volumes with a more collective emphasis, such as the effects of innovative policies on local communities internationally. Cross-disciplinary in scope, these volumes will be suitable for courses in urban politics, community decision-making, urban administration, and related courses, in addition to appealing to mayors, council members, planners, and other policy-makers.

Also in this series

URBAN INNOVATION AND AUTONOMY
Political Implications of Policy Change
Edited by Susan E. Clarke

MANAGING CITIES
IN AUSTERITY

Urban Fiscal Stress in
Ten Western Countries

edited by
Poul Erik Mouritzen

Urban Innovation
Volume 2

SAGE Publications
London · Newbury Park · New Delhi

Chapters 1, 3 and editorial arrangement © Poul Erik Mouritzen 1992
Chapter 2 © Michael Goldsmith 1992
Chapter 4 © Poul Erik Mouritzen and Kurt Houlberg Nielsen 1992
Chapter 5 © Harold Wolman with Michael Goldsmith, Enrico Ercole
    and Pernille Kousgaard 1992
Chapter 6 © Harald Baldersheim 1992
Chapter 7 © Richard Balme 1992
Chapter 8 © Norman Walzer, Warren Jones and Haakon Magnusson
    1992
Chapter 9 © Norman Walzer, Warren Jones, Cecilia Bokenstrand and
    Hakon Magnusson 1992
Chapter 10 © Vincent Hoffmann-Martinot 1992
Chapter 11 © Stephen C. Brooks 1992
Chapter 12 © Poul Erik Mouritzen and Ari Ylönen 1992

First published 1992

 SAGE Publications Ltd
6 Bonhill Street
London EC2A 4PU

SAGE Publications Inc
2455 Teller Road
Newbury Park, California 91320

SAGE Publications India Pvt Ltd
32, M-Block Market
Greater Kailash – I
New Delhi 110 048

British Library Cataloguing in Publication data
Managing cities in austerity: Urban fiscal stress
in ten western countries. – (Urban Innovation)
    I. Mouritzen, Poul Erik   II. Series
    352.1

ISBN 0–8039–8632–7

Library of Congress catalog card number 91-53238

Printed in Great Britain by Biddles Ltd, Guildford, Surrey

Contents

Series Editor's Introduction

Terry Nichols Clark

The Sage series in Urban Innovation emerges from the Fiscal Austerity and Urban Innovation project, which has become the most extensive study of local government in the world. In the USA it includes surveys of local officials in all municipalities of over 25,000 population, nearly 1,000. In some 35 other countries analogous studies are in progress. While project costs exceed $10 million, they have been divided among project teams so that some have participated with quite modest investments. Our goal is to document and analyze the adoption of innovations by local governments, and thus to sharpen the information base of what works, where, and why. The project is unusual if not unique in combining a large scale sophisticated research effort with decentralized data collection, interpretation and policy analysis. The project's potential to help cities provide better services at lower costs has heightened interest by public officials. The wide range of survey items makes the data base a unique source for basic researchers on many related topics. Some data are available to interested researchers via the Interuniversity Consortium for Social and Political Research, Ann Arbor, Michigan. The project remains open to persons interested in participating in different ways, from attending conferences to analyzing the data or publishing in our *Newsletter*, our annual volume, *Research in Urban Policy* (JAI Press), and the Sage series in Urban Innovation.

Books in the Sage series in Urban Innovation may include monographs by a single individual, or collective works on a project theme. We deleted 'fiscal austerity' from the book series since it is not salient to all project participants: urban innovation is. The availability of a comparable core of data from around the world heightens the international interest even of a volume that focuses on a single country, since teams elsewhere may be encouraged to pursue similar work in other national contexts. Volumes may address a topic in depth in one or more English-speaking countries, or compare patterns in two or more non-English-speaking countries.

Volumes for the series are reviewed by the Editorial Board: Terry Nichols Clark, University of Chicago, Chair; Harald Baldersheim, University of Bergen; Susan Clarke, University of Colorado; Gerd-Michael Hellstern, University of Berlin; David Morgan, University of Oklahoma;

Poul Erik Mouritzen, University of Odense; Robert Stein, Rice University. The Board normally meets with Sage staff once a year in Europe, and once in the USA.

BACKGROUND

The project emerged in the summer of 1982. Terry Clark, Richard Bingham and Brett Hawkins had planned to survey the adaptation of 62 US cities to austerity. We circulated a memo summarizing the survey and welcomed suggestions. The response was overwhelming: people across the USA and several other countries volunteered to survey leaders in their areas, covering their own costs. Participants were initially attracted by the opportunity to compare cities near them with others. As it seemed clear that we would cover most of the USA, others volunteered to survey those states not covered. The result was a network of some 26 US teams using a standard methodology to survey local public officials; the teams pooled their data, and then made the information available to the public. The project spread internationally in the same manner.

The Permanent Community Sample (PCS), a national sample of 62 US cities monitored over 20 years, provides a data base and research experience on which the project built. Many questionnaire items, and methodologies for studying urban processes, were derived from the PCS. Fresh data have regularly been made publicly available; a small data file, provided with a self-instruction manual, has been used for teaching at many universities. Several hundred articles and books have used the PCS; the most comprehensive is T. N. Clark and L. C. Ferguson, *City Money: Political Processes, Fiscal Strain and Retrenchment* (New York: Columbia University Press, 1983).

Since it began in 1982, project conferences have been held regularly around the world, often with meetings of larger associations, especially the European Consortium for Political Research in the spring and American Political Science Association in the summer.

Data collection is complete in the US and most European countries; it is still underway in some others. Resurveys to assess changes are under consideration.

FAUI TEAMS WORLDWIDE

The non-US participants are among the leading urban analysts in their respective countries, and in several cases direct major monitoring studies with multi-year budgets including collection of data directly comparable to those in the USA. Gerd-Michael Hellstern, University of Berlin, coordinated the European teams participating in the project. Ed Prantilla coordinated the project on six Asian countries. The survey items are being adapted to different national circumstances while retaining the basic items wherever possible to permit cross-national comparisons.

Participation in the project is relatively open; teams continue to join, especially outside the USA, as they learn of the project and find ways to merge it with their own activities. Austerity is an issue that links the less affluent countries of the world with others, and one with which the less affluent countries have had more experience. Thus they may be able to offer some distinctive lessons.

RESEARCH FOCI

Project participants are free to analyze the data as they like, but past work indicates the range of concerns likely to be addressed. The seven-author statement 'Urban Policy Analysis: A New Research Agenda' in T. N. Clark (ed.), *Urban Policy Analysis* (Beverly Hills: Sage, 1981, Urban Affairs Annual Review, vol. 21, pp. 23–78) outlines several dozen specific hypotheses. Many specific illustrations appear in project publications such as the four volumes of *Research in Urban Policy* completed to date, *Urban Innovations as Response to Urban Fiscal Strain* (Berlin: Verlag Europäische Perspektiven, 1985), edited by Terry Clark, Gerd Michael Hellstern and Guido Martinotti, and several country-specific reports. Over 100 papers have been presented at project conferences and 11 books have been published; these are listed in the *Newsletter*. Some general themes follow.

ISOLATION AND DOCUMENTATION OF INNOVATIVE STRATEGIES

Showcase cities are valuable to demonstrate that new and creative policies can work. Local officials listen more seriously to other local officials showing them how something works than they do to academicians, consultants or national government officials. Specific cases are essential to persuade. But as local officials seldom publicize their innovations, an outside data collection effort can bring significant innovations to more general attention. Questions: What are the strategies that city governments have developed to confront fiscal austerity? How do strategies cluster with one another? Are some more likely to follow others as a function of fiscal austerity? Strategies identified in the survey are being detailed through case studies of individual cities.

IDENTIFICATION OF LOCAL GOVERNMENTS THAT DO AND DO NOT INNOVATE; CLARIFICATION OF POLITICAL FEASIBILITY

One can learn from both failure and success. Local officials often suggest that fiscal management strategies like contracting out, volunteers and privatization are 'politically infeasible'; they may work in Phoenix, but not in Stockholm. Yet why not – specifically? Many factors are hypothesized, and some studied, but up to now much past work is unclear concerning how to make such programs more palatable. The project is distinctive in probing the adoption of innovations, tracing diffusion strategies and sorting out effects of interrelated variables. Interrelations of strategies with changes in revenues and spending are also being probed.

NATIONAL URBAN POLICY ISSUES

In several countries, and especially the USA, fiscal austerity for cities is compounded by reductions in national government funding for local programs. How are cities of different sorts weathering these developments? Scattered evidence suggests that cities are undergoing some of the most dramatic changes in decades. When city officials come together in their own associations, or testify on problems to the media and their national governments, they can pinpoint city-specific problems. Yet they have difficulty specifying how widely problems and solutions are shared across regions or countries. The project can contribute to these national urban policy discussions by monitoring local policies and assessing the distinctiveness of national patterns. Fiscal strain indicators of the sort computed for smaller samples of cities are summarized nationally. Types of retrenchment strategies are being assessed. Effects of national program changes are being investigated, such as stimulation–substitution issues. A several-hundred-page report of key national trends in 12 countries has been published by Poul Erik Mouritzen and Kurt Nielsen, *Handbook of Comparative Urban Fiscal Data* (University of Odense, Denmark, 1988).

CONCLUSION

The project is such a huge undertaking that initial participants doubted its feasibility. It was not planned in advance, but evolved spontaneously as common concerns were recognized. It is a product of distinct austerity in research funding, illustrating concretely that policy analysts can innovate in how they work together. But most of all, it is driven by the dramatic changes in cities around the world, and a concern to understand them so that cities can better adapt to pressures they face. Volumes in the Sage series in Urban Innovation report on these developments.

Preface

Poul Erik Mouritzen

The Fiscal Austerity and Urban Innovation (FAUI) project is the most extensive study of urban fiscal policy-making ever undertaken. Researchers in about 35 countries have conducted surveys with local political and administrative officials using a set of standardized items.

At an early point in the project's history there was a growing interest in comparative studies based on the information collected. At a FAUI conference in Versailles in the fall of 1985 a co-ordinating group was established with the purpose of facilitating such effort. By the summer of 1986 the first comparative book had been initiated by Susan Clarke. Based on a detailed outline, project participants were invited to submit manuscripts which dealt mainly with their own country. The first book in the Sage Urban Innovation series consisted of coherent country analyses for six countries: the USA, France, Holland, Finland, Norway and Denmark (Susan Clarke (ed.), *Urban Innovation and Autonomy: Political Implications of Policy Change*, Sage, 1989).

This book is the second in the series. In contrast to the first, it is organized around topics rather than countries. Ten countries have been selected for the study: Denmark, Norway, Sweden, Finland, West Germany, the UK, France, Italy, Canada and the USA. To the extent that data are available, we analyze topics such as fiscal stress, political environments and fiscal strategies across these countries, trying to detect and explain differences and similarities. We draw on aggregate national data as well as city-level data.

These analyses have depended on extensive commitments and loyalty to the group on the part of the individual participant. Great burdens were often laid on individuals requiring them to do analyses over and over again and to collect additional information for hundreds of cities. In this process, disappointments were few while the belief that scholars can be asked to do extensive work of high quality without immediately visible benefits was strengthened.

At the outset the participants were organized into four groups, each directed by a topic co-ordinator. Harold Wolman directed the work on central–local relations, Harald Baldersheim the work on political environments; Norman Walzer was in charge of the fiscal strategy group, while Poul Erik Mouritzen directed the studies on fiscal stress. This organization enabled each participant to take on two expert roles: expert on a certain topic and expert on their own country.

The preparation of the book was greatly facilitated by frequent meetings. The annual meetings of the European Consortium for Political Research and the American Political Science Association were used as occasions for short project meetings. Two longer meetings, specifically set up for the work on this book, were very important. Harald Baldersheim and the LOS Institute in Bergen were the hosts of a 10-day data confrontation seminar in September 1987, where mornings were used for intensive presentations by the topic groups of their ideas and afternoons for the first data analyses. Enrico Ercole organized a follow-up meeting in Bologna in April 1988 where new ideas were exchanged and the manuscript finally took form. The project participants owe great thanks to Baldersheim and Ercole for their invaluable work in setting up these well organized conferences.

Finally, the work on the book was greatly facilitated by the Fulbright Foundation which supported the volume editor's visit at the University of Colorado, Boulder, in 1988–89.

The research group wish to thank Hanne Bille and Arne Vesth Pedersen who typed and corrected the final draft for publication.

The Contributors

Harald Baldersheim has been professor of public administration at the University of Bergen, Norway, since 1977. He is currently affiliated with the Norwegian Research Centre in Organization and Management, Bergen. He was responsible for the co-ordination of evaluation research on the Norwegian 'free commune' experiments and the Norwegian Fiscal Austerity and Urban Innovation Project. He has published several books and articles on urban policy.

Richard Balme is chargé de recherche for the Fondation Nationale des Sciences Politiques. He works at the Centre d'Etude et de Recherche sur la Vie Locale in Bordeaux, France. From November 1986 to June 1988, he was visiting scholar at the University of Chicago, contributing to the Comparative Analysis of the FAUI project. He has published several articles about political participation, collective action and local policies.

Cecilia Bokenstrand is a PhD candidate in political science at the University of Gothenburg, Sweden. As a research assistant, she has been working in the Swedish Fiscal Austerity Project since 1986 and she directed the Swedish Fiscal Austerity mail survey with Professor Lars Strömberg.

Stephen C. Brooks is associate professor of political science at the University of Akron, Ohio. He was co-director of the Ohio Fiscal Austerity and Urban Innovation Project and has studied urban fiscal stress in Britain. He has written on both methodological and substantive issues raised by project data.

Terry Nichols Clark has been president of the Research Committee on Community Research of the International Sociological Association for almost 20 years; its activities led to the Fiscal Austerity and Urban Innovation Project in 1983. As the Project's international co-ordinator, he edits a *Newsletter* and the JAI annual, *Research in Urban Policy*. He has taught at Columbia, Harvard, Yale, UCLA and the Sorbonne, and is now Professor of Sociology at the University of Chicago. He has worked at the Brookings Institution, the Urban Institute, the US Conference of Mayors, and the Department of Housing and Urban Development. He has consulted with many cities on fiscal management questions and has published over 100 books and articles on urban policy issues.

Enrico Ercole is a doctoral student in sociology at the University of Turin, Italy. He participated in research projects on urban development, urban politics and urban culture. Currently co-ordinator of the Italian Fiscal Austerity and Urban Innovation Project under the supervision of Prof. Guido Martinotti, University of Pavia, Italy, he is author of several articles on urban problems in Italy.

Michael Goldsmith is a pro-vice-chancellor and chairman of the Department of Politics and Contemporary History at the University of Salford. He has written and researched extensively in the comparative urban politics and local government fields, and his recent publications include *New Research in*

Central and Local Government Relations (Gower, 1986); and a volume with Ed Page: *Central and Local Government Relations: A Comparative Analysis* (Sage, 1987).

Vincent Hoffmann-Martinot studied at the University of Bordeaux, where he prepared a doctorate on the financial autonomy of West German local authorities. He is currently a researcher at the Centre National de la Recherche Scientifique in Bordeaux. He is the author of *Finances et pouvoir local: l'expérience Allemande* and has written widely on questions of local politics and policies in France and West Germany.

Warren Jones is associate professor of economics at Western Illinois University. His research has focused on social security, user charges and municipal expenditures. He has also consulted for the Governer's Task Force on Rural Illinois.

Pernille Kousgaard is a research fellow at the University of Salford. She has been responsible for the collection of the English local government financial statistics for the Fiscal Austerity and Urban Innovation Project since 1987. She is the co-author of several articles on local government finance.

Haakon Magnusson is associate professor at the Department of Political Science, Lund, Sweden. He directed the study on local budgeting as part of the second Swedish Local Government Research Project in the late 1970s and was a co-director of the FAUI project in Sweden. He has published several books and reports on local government budgeting and fiscal policies.

Poul Erik Mouritzen is associate professor and chairman of the Department of Commercial Law and Political Science, University of Odense, Denmark. He has been a visiting scholar at the University of California, Irvine, under the auspices of the American Council of Learned Societies. In 1987/88 he spent a year at the University of Colorado, Boulder, as a Fulbright research fellow. From 1980 to 1985 he directed the Danish Fiscal Austerity Project. He has published articles on fiscal austerity, citizens' preferences and local government fiscal policy-making.

Kurt Houlberg Nielsen was a research assistant on the Danish Fiscal Austerity project. He was responsible for the collection of comparative aggregate data for the international project. He now works as a consultant for local government in Denmark.

Norman Walzer is professor and chairman of the Department of Economics at Western Illinois University. He co-ordinated the Illinois section of the Fiscal Austerity and Urban Innovation survey in the USA and has worked with the data since then. He has published extensively on local public finance issues including six books and numerous articles in the *Review of Economics and Statistics, Public Finances, Public Choice, Land Economics,* and the *National Tax Journal.*

Harold Wolman is professor of political science and research scholar in the College of Urban, Labor, and Metropolitan Affairs at Wayne State University in Detroit, Michigan. He previously served as senior research associate at the Urban Institute. Dr Wolman is the author of a variety of articles concerned with urban fiscal stress, many of them in cross-national context.

Ari Ylönen is director of the Research Institute for Social Sciences at the University of Tampere, Finland. He is responsible for the Finnish contribution for international comparisons in the Fiscal Austerity and Urban Innovation Project. He has been teaching sociology as acting professor at the University of Tampere, and has published several articles and research reports on urban sociology and housing.

PART I
THE CONTEXT

1
Introduction

Poul Erik Mouritzen

FARUM ON THE BRINK OF COLLAPSE

In the fall of 1982 time was running out for Mayor Gøsta Gustavsson. For many years he had been the unquestioned leader of the city of Farum, a suburban middle-sized municipality situated at the outskirts of the Copenhagen metropolitan area. No one in the Social Democratic Party had questioned Gustavsson's ability to lead the city and the party – at least until now. Despite fairly favorable external conditions, however, city finances were in a mess. Lavish spending and uncontrolled investments had been followed by a gradual erosion of city finances. Debts were sky-high, the city tax rate was among the highest in the country, and the bank account was in the red. During those months in 1982 the city was unable to pay salaries to its employees. The Mayor's response was to postpone payment of collected taxes to the National Tax Authority. Suddenly, Farum attracted national attention.

Although under heavy pressure from within his party, Gøsta Gustavsson managed to survive. However, a year later he shifted his party affiliation to the Liberal Party, from which point he managed to survive as city mayor until the 1985 local election.

Four years later, on election night, 21 November 1989, Mayor Peter Brixtofte could await the returns with confidence. Previously a young member of Parliament for the Liberal Party, he had taken over the mayorship in Farum in 1985. Years of innovative policy change had followed. Services had been trimmed drastically, the administrative staff had been reduced (from 210 in 1988 to 168 in 1990, with a planned target of 133 in 1992), and, most important, for five consecutive years the local tax rate had been reduced, from 20.7 percent in 1985 to 19 percent in 1989 (later to 17.9 percent in 1991), well below the national average. Net interest payments and retirement on debts were again close to the average for the metropolitan area. These policy changes were all the more impressive as they had been implemented in a period of great external stress, mostly caused by major cutbacks in grants from central to local governments and a comparatively large influx of refugees and immigrants from Third World countries. Farum was again attracting national attention, this time as a model city for other municipalities in terms of fiscal management.

Two hours after the polls had closed on November 21, Mayor Brixtofte could reap the benefits from his fiscal policies, and, what was perhaps more important, from his status as one of the most effective local leaders in the country. His party gained an impressive 53 percent of the votes in Farum, an increase from 22 percent in the 1985 election. Nine more seats were won in the City Council, where the Liberal Party now had an absolute majority (13 out of 21 seats). Peter Brixtofte was sure of four more years in the mayor's seat.

THE OBJECT OF THIS VOLUME:
THE THOUSAND OTHER FARUMS

In this book thousands of Farums in ten Western countries are studied. Most of them are anonymous. Many have been exposed to severe fiscal difficulties, like Farum in the beginning of the 1980s. Important examples are Liverpool, Oslo and New York City. Some have been able to turn the tide, like Farum in the 1980s. The Farum case contains many of the elements that make the study of urban policy-making exciting, elements that will be studied more systematically in this volume.

First of all, Farum's revival from near bankruptcy is indicative of the importance of political leadership. Cities and leaders can survive, even under severe external pressure – and Farum was under severe pressure during the 1980s. A lot of Danish municipalities, most of them starting the decade from a much better fiscal position than Farum, were close to bankruptcy toward the end of the decade. Leaders do make a difference.

The Farum case seems to confirm the popularly held belief that political parties are important. Social Democrats are spenders who do not care too much about the taxpayer's money and future generations who will have to pay the bill; however, they give you a good service. Bourgeois party leaders represent the taxpayer, are enemies of bureaucracy, and are able to produce more for less. Are these contentions just myths, or are they grounded in reality? We try to address this question in several of the following chapters.

Shifts can take place rapidly. Like most Danish municipalities, Farum suffered from fiscal stress caused mainly by cutbacks in grants to local governments. From a period of prosperity and drastic improvements in services (in real terms, current expenditures increased by 20 percent from 1978 to 1982), citizens as well as political leaders had to get used to zero or negative growth in the subsequent years.

Central government plays a vital role for the well-being of local governments. Again, popularly held beliefs find substantiation in the Farum case. In 1982 the Social Democratic administration gave up and was followed by a Conservative coalition government. Two months later a revival program was launched, in which reductions in the grants to local governments were a major element. Are left governments big spenders and conservative governments more critical of a large public sector, as common knowledge teaches us? This question is addressed more systematically in the book.

Finally, the Farum case fits into a much discussed issue of the 1980s, the role of the welfare state. Mayor Brixtofte of Farum did not fundamentally question the welfare state. By 1990 families were still paying less than a third of total costs for day care, the average size of classes in Farum's school system was 18, and libraries could be used at no cost. (By many foreign standards, Brixtofte is a socialist!) However, he started a process of slowly and marginally reshaping the welfare state. Probably more important, the fact that people support him can be taken as an indication of changes in the voters' perception of and attitudes towards the welfare state. Brixtofte in some respects resembles the so-called New Fiscal Populists, leaders who are fiscally conservative but socially liberal, and who appeal to the individual voters with no commitment to special interests.

OVERVIEW OF THE VOLUME

In Chapter 2 by *Goldsmith* the 10 selected countries are described with respect to the main features of their local government systems. Goldsmith highlights the electoral systems, the functions performed by local governments, the finance systems and the role of political parties, and proposes a distinction between three models of local government: the Scandinavian (or northern European), the southern European and the North American models.

Part II of the book focuses on the causes of urban fiscal crises. *Mouritzen* first sets out to discuss the various components of the fiscal crisis (Chapter 3). It is argued that fiscal austerity has objective as well as strong subjective elements. The notion of fiscal crisis is deeply rooted in the political process and performs important political and symbolic functions for actors involved in local politics. A concept that allows for measurement across countries is proposed and an operational definition is established. This concept of fiscal slack and its opposite, fiscal stress, is used in the remainder of the book.

Was there a fiscal crisis in the 1980s? That is the question *Mouritzen and Nielsen* try to answer in Chapter 4. The answer to this question is yes – for some countries. However, a number of countries, notably Norway and Finland, have since the end of the 1970s, experienced extremely favorable conditions for local governments, with yearly increases in revenue.

In Chapter 4 the cause of fiscal crisis is decomposed into various contributing factors. It is shown that by far the most important factor defining the fiscal well-being of localities is the grant policies pursued by central governments. In Chapter 5, *Wolman, Goldsmith, Ercole and Kousgaard* present a number of detailed analyses of the causes and consequences of government grant policies. One of their questions, which is addressed for the first time in a comparative context, is: to what extent is the distribution of grants to local governments guided by partisan politics or notions of 'objectivity'?

In Part III the two contributions focus on the political environments of local leaders. In Chapter 6 *Baldersheim* studies the effects of a fiscal crisis on the policy maps of political leaders, that is, on their perceptions of problems and challenges faced by their municipalities, on their perceptions of the political environment, on their success in implementing their own spending preferences and on their leadership style. Some of the elements of these propositions are derived from certain system characteristics and leadership patterns which distinguish the three countries under investigation, Norway, France and the USA.

Balme in Chapter 7 investigates how leaders' fiscal preferences vary as a function of present policies, responsiveness, ideology and government structure. These preferences are compared with the preferences of citizens as perceived by the politicians, and differences – or policy distance – are described for five countries.

Part IV focuses on the strategies pursued by local political leaders. Chapters 8 and 9 are to some extent based on the same set of data; however, they are analyzed from two different perspectives.

The main theme of Chapter 8, by *Walzer, Jones and Magnusson*, is the relationship between fiscal slack and policy responses. They start out with national (aggregated) data and end up with a test based on city-level data covering five countries. Their main finding corresponds to what we normally find in the policy output literature: that party politics are relatively unimportant, whereas the degree of fiscal slack is a major determinant of growth in local spending.

Chapter 9, by *Walzer, Jones, Bokenstrand and Magnusson*, is organized by strategy. Based on aggregate statistical data as well as information obtained in surveys, the authors compare the 10 countries with respect to the extent to which different strategies have been applied, such as reductions in expenditure (capital, current and salaries), increasing reliance on user charges and increases in taxes. They systematically compare the perceptions of leaders (obtained via surveys) with actual behavior (data from traditional statistical sources). From a methodological point of view, their findings are rather optimistic as perceptions seem to reflect actual fiscal behavior rather precisely.

One particular set of strategies is the subject of *Hoffmann-Martinot*'s study in Chapter 10. The number of municipal employees, their organizations and their activities define important aspects of the political environment of local politicians. The first part of the chapter discusses how these environments differ between countries, while the last part investigates cross-national differences in the application of strategies involving municipal personnel, such as the size of the workforce, compensation levels and privatization.

A set of strategies, normally invisible to the average citizen, concerns the application of modern fiscal management techniques. *Brooks* in Chapter 11 discusses the application of such techniques as revenue forecasting, fiscal information systems and performance measures. The main question in this chapter is whether management innovation is related to fiscal crisis and in what way.

In Part V *Mouritzen and Ylönen* (Chapter 12) consider some of the main findings in relation to the theoretical framework and main propositions. A central concern is the extent to which urban fiscal stress affects fiscal policy-making. Are there distinctly different patterns of policy-making and different policy outputs in situations of stress relative to situations of slack? Or are differences explained mainly by institutional contexts, that is, by country? Or are partisan politics the major determining factor?

The last part of the volume is a Technical Appendix. This includes a broad overview of the data used in the study and covers some of the problems of standardization. Finally, the FAUI projects in each country are described in detail, with respect to time of the surveys, samples, response rates, categories of actors surveyed, etc.

All the local government data (whether statistical or survey data) used for this book refer to municipalities. We use the terms municipality and local government as synonyms.

2

The Structure of Local Government

Michael Goldsmith

The purpose of this chapter is to provide a contextual description of the main features of the local government systems in each of the countries that are the focus of the study. The intention is not to provide a detailed description of each country's local government structure, but more simply to draw attention to the main elements of structure, function, finance and politics of the different systems, together with an attempt to highlight the most important points of similarity and difference.

Such a description will help the reader both to understand the context within which each country has operated in recent years, and to interpret the comparative analysis of the different aspects of fiscal austerity and urban innovation undertaken in subsequent chapters.

First, a word of caution is necessary. All efforts at comparison inevitably lose some of the detail. In searching for common denominators, we run the risk of forgetting some of the important differences, and nowhere is this more true than when dealing with the world of local government. Thus, in no two countries is the allocation of functions or competencies going to be similar; indeed, simply to say, for example, that education or public housing is a municipal function in Denmark and France is not to say that municipalities in Denmark and France have similar education or housing responsibilities.

It is important also to remember that here we are dealing with local government systems and their behavior as a whole, and not with individual municipalities or cities. Thus, for example, to say that British local government has been going through a period of fiscal stress while Norway has not, is not to say that Canterbury has been fiscally stressed or that Bergen has not faced fiscal problems, although where possible analysis of city data is included.

ELECTORAL SYSTEMS

With these provisos in mind, what can we say about the local government systems of the different countries? First, they are democratic, in the sense that they are based on some form of *direct* election. Even so, as Table

Table 2.1   *Method of election and constituency in municipal authorities*

	Method of election	Constituency
Denmark	PR	At large
Norway	PR	At large
Sweden	PR	At large
Finland	PR	At large
Germany	Mixed	Mixed
UK	Simple majority	Ward
France	Mixed, complex	At large
Italy	PR[1]	At large
Canada	Simple majority	Ward
USA	Simple majority	Mixed

PR Proportional representation.
[1] In cities of over 5,000 population.

2.1 shows, both the method of election and the basis on which council members are elected varies from country to country, and even within countries. Clearly, as we would expect, those countries that use an electoral system based on proportional representation (PR) all have at-large elections, even if the PR system itself may vary. Similarly, most of the countries using simple majority systems also elect people on a ward, or small geographical areal, basis. Even so, two of the countries with which we are concerned – West Germany and the USA (both federal systems) – elect people on a mixed basis.

But the basis on which people are elected to local councils has important consequences both for the *representative* nature of the local system and for the possible *mobilization of bias* in policy responses to fiscal austerity and urban innovation. Ward-based systems may mean that strategies designed to cut services may be more difficult to adopt, since the areal and social consequences of cuts may be more readily apparent and more easily opposed by local politicians. Similarly, some PR systems, such as list systems, may produce distortions in the partisan nature of municipalities, whereas others may ensure widespread minority party representation, perhaps leading to minority government and the development of more consensual styles of local politics. For example, such a style of local politics/government has been common in Scandinavian countries in the post-war period, whereas the rise of third parties in Britain in recent years has thrown many local councils into disarray; accustomed to the normality of majority government, they now find themselves 'hung councils', with no party in majority and occasionally no one willing to take the leadership role.

As far as the frequency of elections is concerned, in the great majority of Western nations all councillors are elected at the same time, although Britain and the USA have staggered terms of office. In Britain and Sweden elections are every three years, in Denmark, Finland and Norway every four, and in France every five or six years. Turnout in local elections, however, varies, from an average of 25 percent in the USA to one of 90 percent in Sweden; Italy comes next (83 percent), with Denmark, Finland, France and Norway having average turnouts of over 70 percent, then Britain at 40 percent and Canada at around 33 percent. It is difficult to judge what these differences mean for local politics in practice, except perhaps to suggest that high turnouts are more likely to result in a high turnover of elected representatives.

The size of local councils, together with the number of people for which they are responsible, also varies from country to country. In general, and excluding capital cities, West European countries favor large councils, with the exceptions of France and Norway, where most councils have fewer than 50 members. The number of cities and large towns (or the proportion of urban population) is likely to be an indicator of the number of large councils, but even among Western European countries this can produce some odd results. For example, English municipalities average populations of over 120,000: the French just over 1,300, Sweden about 30,000 as against Norway's 8,800 and Denmark's 18,000. The US municipalities average just over 14,000, whereas in West Germany the figure is just under 3,000. Not only do these figures reflect the sizes and population densities of different countries, they also reflect ideas about the appropriate scale of operation of local government. For example, the size of British local authorities, and perhaps those in Scandinavian countries, more accurately reflect ideas about efficient and economic service provision, whereas the size of American or French municipalities may reflect ideas about the value of community and local democracy.

In this context, it is important to consider the ratio of citizens to local elected representatives. Though it is difficult to obtain accurate figures, the ratio is somewhere between 1 : 250 and 1 : 450 in most Western nations: in Britain the ratio is approximately one councillor to every 1,800 citizens (Goldsmith and Newton, 1986: 140). Differential ratios of this order will obviously reflect how easily elected representatives and their fellow citizens can interact with each other, and the extent to which the former can accurately reflect the preferences of the latter, something also affected by the basis on which councillors are elected.

THE FUNCTIONS OF LOCAL GOVERNMENT

Comparing what local governments do in most countries is notoriously difficult: describing the allocation of functions to local levels is even more so. Though collected in 1986 for other purposes, Figure 2.1 gives

Functional Classification	Denmark	Norway	Sweden	Finland	Germany	UK	France	Italy	Canada	USA
Refuse collection & disposal	L	L	L	L	L	L	L	L	L	L
Slaughterhouses	L	L	L	L	L	L	L	L	N/A	N/A
Theatres, concerts	R	L	L	L, S	R, L	L	L		L, P	L
Museums, art galleries, libraries	R, L	L	L	L, S	R, L	L	L, D	R, L	L, P	L
Parks & open spaces	L	L	L	L	L	L	L, R	R, L	L, P	L
Sports & leisure pursuits	L	L	L	L, S	R, L	L	L	R, L	L, P	L
Roads	R, L	R, L	R, L	L	R, L	L	L, D	R, P, L	L, P	L
Urban road transport	L	L	L	L	R, L	L	L, D	P, L	N/A	L*
Ports		L	R, L	L	R	L		L	N/A	N/A
Airports	L	L	R, L	L	L	L	L	L	N/A	N/A
District heating	L	L		L		L		R, L	L	L
Water supply			N/A	N/A	L	L	L	P	N/A	N/A
Agriculture, forestry, fishing, hunting	L	L		R, L	L	L, D	R, P, L		P	N/A
Electricity		L	L	N/A	R, L	L	L	R, L	P	N/A
Commerce			L		R, L	L	L, D	R	N/A	N/A
Tourism		L	R, L	S	R	L	R, D	R	P	N/A
Financial assistance to local authorities				S	R, L	L	L	L	L, P	L, S
Security, police	L	L		L	L	L, D	L	L	L*	
Fire protection			L	S	R	L		L	N/A	N/A
Justice						L		R, L	L, S	L, S
Pre-school education	L	L	L	L	L	L	L	R, L	L, P	L, S
Primary & secondary education	R, L	R, L	R, L	L	R, L	L	L	R, L	L, P	S
Vocational & technical training		R, L	R, L	S	L	L	L	R, L	P	S
Higher education					R	L		R	P	S
Adult education	L	R, L	R, L	L	R, L	L	L, D	R	P	S
Hospitals & convalescent homes	R	R, L	R	R	R, L	L	L	R, P, L	L, P	L*
Personal health	R, L	L	L	R, L	R, L	L	L, D	R, P, L	L, P	L*
Family welfare services	L	L	L	L	L	L	L	R, L	L, P	L
Welfare homes	L	L	L		R, L	L	L	R, L	L, P	L
Housing	L	L	L	L	R, L	L	L	R, L	L, P	L
Town planning	L	L	L	L	R, L	L	L	R, L	L	L

Figure 2.1 *The functions of local government (HMSO, 1986: Fig. K2 (often more than one municipality), and other FAUI project surveys)*

D = department   L = local   P = province   R = region   S = state some idea of the way in which functions, broadly defined, were allocated between different levels of sub-national government in different countries at that time.

Even a representation such as that presented in the figure can be dangerously misleading, however, if only because the 'functions' are little more than labels and tell us little about the actual content involved. Breaking any one function down into meaningful categories in terms of types of capital and current expenditure activities quickly produces an even more bewildering picture than that presented here. But the figure is misleading in another sense; for it is little more than a snapshot taken at a particular point in time, while in practice the allocation of activities is dynamic and constantly changing. Municipalities take on or develop new functions or give up others; intermediate levels develop new localized agencies to undertake some new function on an ad hoc special-purpose basis, while central governments may review and redistribute functions between different sub-national levels from time to time.

In this context, the extent to which there is considerable territorial, programmatic and functional consolidation of both local government areas and services is an important factor, even though local services are likely to be provided by a variety of agencies in all countries. The USA, for example, is characterized by the widespread fragmentation of its local government system, in areal, functional and fiscal terms. In any urban area, local services are likely to be provided by a mixture of state and federal agencies; a plethora of ad hoc single-purpose bodies (both elected and appointed, with or without taxing and charging powers), as well as by the municipal governments themselves. In other words, the extent of local consolidation in the USA is very low, and appears likely to remain so: local government is thus fragmented, and this fragmentation is a stable condition or feature of the system.

By contrast, the Scandinavian countries have local government systems which are highly consolidated in territorial and functional terms. The local governments of Norway, Sweden, Denmark and Finland are wide-ranging multi-purpose service-providing agencies, and are likely to remain so.

In this context, Britain and France provide contrasting examples of consolidation. Britain began the 1980s with a local government system with a reasonably high level of consolidation, but as the decade proceeded a series of Conservative governments led by Mrs Thatcher increasingly fragmented the system, by such actions as the abolition of the GLC and the metropolitan counties, the creation of various local development agencies and corporations, and other changes in services such as housing and education. By contrast, the French reforms of 1987 and subsequently during the Mitterrand presidency have seen both a decentralization of functions down to intermediate tiers and some consolidation in territorial terms. Thus, unlike the more stable Scandinavian countries and the USA,

Table 2.2   *Size of local government in relation to total public sector*

	Share of public expenditure[1] %	Share of public employment[2] %
Denmark	46	70
Norway	38	64
Sweden	41	54
Finland	43	NA
Germany	18	15
UK	30	38
France	18	11
Italy	29	12
Canada	19	NA
USA	31	50

[1] All local government expenditure as a percentage of total public expenditure in 1984 (Germany in 1983).

[2] Local government employment as a percentage of total public employment in 1980 (Sweden, Germany and USA in 1981; France in 1982).

Sources: (column 1) (Mouritzen and Nielsen, 1988: 44); for Finland: *Finnish Statistical Yearbook*; (column 2) Page and Goldsmith (1987: 157).

British local government is becoming increasingly fragmented, while that of France is becoming more consolidated.

Trends such as these make the measurement of the extent and nature of local government activity in different countries increasingly fraught. Two measures sometimes used are the relative shares of public expenditure and employment taken up by the local government sector. While likely to be open to detailed questioning for any particular country, such measures are sufficiently accurate for us to be able to discriminate usefully between countries and/or groups of countries. Table 2.2 presents these data for the countries included in our sample.

This table, again, brings out the distinction between the countries in northern Europe, including Britain, and those in southern Europe, and permits a further distinction to be made between unitary and federal systems of government. Thus, the northern European countries, all of which have been essentially social democratic welfare states for most if not all of the post-war period, have local government systems that encompass more of both public expenditure and employment than do their southern counterparts, where local government has largely remained small-scale and based more on patronage or clientelistic relationships. The federal systems of West Germany and Canada also involve relatively small shares of public employment, though in the USA local government in all its forms accounts for half of all public employment. Nevertheless, in public expenditure terms, all three federal systems do not account for large shares of public expenditure, in this case because the intermediate tier has been largely responsible for whatever welfare state services are provided.

THE FINANCE OF LOCAL GOVERNMENT

Local government in most countries is financed by some permutation of local taxes, fees/user charges and central government grants, as well as by borrowing, mainly if not exclusively for capital projects. The result is that the detailed finance of local government in any one country is likely to be somewhat complex. For example, in terms of local taxes, the countries with which we are concerned use a wide range of different types of local taxes, all of which are likely to have different advantages and disadvantages when it comes to raising taxes when a municipality is attempting to cope with fiscal stress. Local government in some countries, such as Italy and the Scandinavian countries, have a wide range of taxes available to them, while others, such as Britain and Canada, are limited to only one, somewhat unbuoyant, tax, namely the property tax. Some countries depend more on a local income tax, others on sales taxes of some sort or other. The elasticity of taxes is dependent in part on their buoyancy during periods of inflation, or else on their visibility, that is, on the extent to which people are aware of paying them. Thus, taxes such as income tax, purchase or sales taxes and those on alcohol and tobacco are less visible than other taxes, since they are either hidden in the price of goods and services or deducted at source. Property taxes are more visible, since they are generally payable out of income and on demand, or in relatively large amounts: they are also generally less buoyant than other taxes, since the basis on which they are calculated often does not change very much over time. Thus, countries like Norway, Sweden and Denmark have less visible local tax systems (albeit generally higher rates of tax), while countries like Canada, West Germany and the USA have more visible property-based systems.

Taxes on business and industry are another possible source of local revenue. Britain's businesses used to pay the full rate of property tax, while local residents' property taxes were subsidized currently by government grant. This was one reason why the Conservative government chose to introduce the community charge, payable by all local residents at a fixed per capita rate, with businesses paying a nationally determined tax on the value of their property. France provides an example of a country with an interesting mix of taxes involving one on land and buildings, one on occupied property, and the *taxe professionelle* (a tax on business which is related to the size of the payroll). Again, the elasticity of such taxes on business will depend on how far they can be passed on to the consumer

Table 2.3 *Percentage distribution of total municipal revenues, 1984*

Source of revenue	Dk	No	Sw	Fi	G[3]	UK	Fr	It	Ca	USA
Individual income tax	45	42	42	40[1]	15	0	0	0	0	6
Corporate income tax	1	4	0		15	0	14	0	4	0
Property tax	2	3	0	0	5	29	8	0	28	20
Other taxes	0	7	0	0	0	0	14	10	7	15
General grants	11	5	4	21[2]	28[2]	24	21	53[2]	6	8
Conditional grants	23	28	22			20	15		43	23
Fees and user charges	21	9	19	15	25	13	9	9	10	14
Other revenues	−4	2	13	24	13	15	20	29	5	13
Total (%)	99	100	100	100	101	101	101	101	103	99

[1] It is not possible to distinguish between individual and corporate income tax.
[2] It is not possible to distinguish between general and conditional grants.
[3] The German table includes all local governments.

Source: Mouritzen and Nielsen (1988: 53–151, cf. country index tables).

and the extent to which the business interest feels it legitimate to pay them. In this context, it is interesting to compare the relatively high level of complaint made by British businesspeople about what they see as the heavy burden of property taxes (which they argue is a significant cost in competitive terms) with the relatively low level of complaint by French businesspeople.

What is important about local taxation in the end is the proportion of local resources that comes from local taxation. Where that proportion is relatively low (as in Holland, where local taxes account for little more than 5 percent of local municipal income), the burden and visibility of local taxation are relatively small. Italy is another example: local taxes account for less than 10 percent of municipal resources (cf. Table 2.3). Conversely, where local taxes provide a significant share of local resources (for example, in the Scandinavian countries it approaches 50 percent of total local revenues), their burden is somewhat higher. By contrast, in Canada, West Germany and the USA, local taxes account for around one-third of local revenues, but the dependence of local government on the property tax in these countries probably makes their local tax burdens just as heavy as the indirect taxes used in Scandinavia. In all these countries, increasing local taxes is not likely to be a preferred strategy when it comes to dealing with fiscal stress.

In this context it is useful to consider the changing pattern of tax effort in our various countries. Of course, tax effort is related to tax need, that is to say, to the extent to which different local systems need to use taxes as a source of income. Thus, some countries, of which Norway would be an example, had buoyant economies over much of our review period between

1978 and 1984, as a result of which total tax effort actually declined by some 6 percent, while in Canada it declined by over 10 percent. But of course there may also be some unwillingness to increase taxes because of their visibility: the Canadian figures may include an element of this factor. The biggest increase in tax effort is recorded in Italy (60 percent), with other countries recording an increase of between 7 and 20 percent (Britain, Denmark, West Germany). Norway, Sweden, Finland and the USA report slightly reduced tax efforts of between 2 and 8 percent. (All figures from Mouritzen and Nielsen, 1988: 30.)

If the different types of local taxation systems are complex, then grant systems are even more so. The simplest, and in many ways most important, distinction is between those countries that essentially use general or block grants to finance local government, and those countries that use specific or categorical grants. The distinction between them is that the first type of grant system is generally unrestricted, in that the money is allocated to municipalities and they are 'free' to use it for any legal purpose. By contrast, specific or categorical grants come with conditions attached and are not normally designed to meet the full cost of providing a particular service.

The Netherlands is the best example of a country where local government is heavily dependent on central grants for the necessary resources to finance local services: in 1984 something like 81 percent of municipal resources came from such grants. Italy is another example, with over 66 percent of local resources coming from grants. At the other extreme, in Sweden, France and West Germany localities get less than 30 percent of their resources from centrally provided grants. Britain, Canada and Norway all take around 40 percent of their local resources in the form of central grants, with the USA and Denmark taking around 33 percent.

But it is also the balance between general and specific grants that is of interest. Of the countries with which we are concerned here, in both Finland and Britain local governments receive over 20 percent of all their revenues in the form of general grants: for all the other countries for which we have data, the figure is less than 10 percent (Denmark, Norway, Sweden and the USA). By contrast, in the latter group local governments take over 20 percent of their resources in the form of specific conditional grants, something that is likely to restrict their discretion to follow their own designs quite considerably. In this context Britain provides an interesting case, in that local authorities acquire something like 20 percent of their resources in the form of specific grants, a trend that has been growing throughout the 1980s, reflecting central government's desire to gain greater control over the way in which British municipalities spend their money.

As far as fees and user charges are concerned, these have grown in importance in most countries over the last decade. Increasing such charges is a well-known and well-documented strategy (Clark and Ferguson, 1983) for local authorities attempting to raise revenues in the face of fiscal stress. The ability to do so, of course, depends on a variety of factors, such as the range of services provided by local governments in a particular country; the legal position of such charges, and cultural attitudes to their use. For example, in those countries where local government provides relatively few services, the ability to charge for them is likely to be limited. Data from our countries would seem to support this view: for example, in both France and Italy, where the range of local services is relatively narrow, the share of total revenue raised from fees and charges is around 9 percent. But cultural or legal factors may be equally important. Norwegian municipalities provide a wide range of services, but they also raise only 9 percent of their income from fees and charges.

By contrast, West Germany (25 percent), Denmark (21 percent) and Sweden (19 percent) all raised considerable resources from fees and charges in 1984. Other countries, such as the USA and Finland (15 percent), Britain (13 percent) and Canada (10 percent), raise considerably less. Notwithstanding the range of differences that these figures reveal, however, it is clear that, compared with grants and taxes, fees and charges are a less important source of local revenue.

PARTISAN AND POLITICAL DIFFERENCES

Willingness to use resources and to provide a particular mix of local goods and services also reflects important political and partisan differences at the local level. In other words, the discretion of local government in a particular country is a reflection not only of its constitutional position, but also of the way in which that discretion is interpreted by local politicians. In terms of the constitutional position of local governments, it is useful to draw a distinction between those countries in which local government either formally has or is presumed to have some level of general competence, as in the Scandinavian countries, France, West Germany, Italy and parts of the USA, and those countries in which local government has only specific competencies – in our group, Britain and Canada. Whether this formal constitutional position makes any real difference to the autonomy of local governments in different countries is, however, more debatable, since, as Page and Goldsmith (1987) show, discretion is a tricky variable to handle and analyze at the best of times. Higher levels of government almost everywhere seem to find ways and means of restricting what sub-national governments actually do – for example by making some functions mandatory, by some system of administrative supervision of local services and agencies, through the use of specific grants, or whatever.

What is likely to be more pervasive is the nature of partisan differences at the local level, and the degree of intensity that exists between local partisan differences. The local councils of most Western nations have become increasingly party-political over the last two decades or so, and the majority now have well developed competitive party systems, especially in the case of large cities and urban areas. It is important, however, to distinguish between two aspects of party politics. The first, partisanship, refers to the extent to which party organization, especially national, permeates local politics, from elections to the determinants of local policy. The second, intensity, refers to the style and temperature of political conflict, both between different parties and between parties at different levels of state institutions. The two are not necessarily highly connected. Thus, Sweden and Finland, for example, both have a highly developed sense of partisanship but a low level of intensity, especially in the light of the dominant consensual style of politics in both countries. The same appears to be true for Norway, Denmark and West Germany, though there are some signs that intensity is increasing in both Norway and Denmark.

By contrast, partisanship and intensity are features of local politics in both Britain and Italy. In both countries parties permeate local levels, and the intensity of conflict between different parties and different levels is acute (Sanantonio, 1987; Rhodes, 1988).

Our two North American countries – the USA and Canada – represent examples where there is a relatively low level of partisanship within the systems of local government. Despite the well-known examples of partisan cities in America, well over 60 percent of cities with populations of above 5,000 are formally non-partisan, which is to say that candidates for political office may not by law identify themselves by political party on the ballot paper. Non-partisanship of this type is a feature of many midwest and western cities in the USA. Despite this non-partisan regulation, however, there are many cities where parties operate covertly: in Chicago for example, local politics are formally non-partisan, although this fact would not be well known to most Democrats there. Canada is an outstanding example of non-partisanship, a country where, with few exceptions, attempts by national parties to enter the local political arena have generally met with failure (Masson and Anderson, 1972; Magnusson and Sancton, 1983). It is more likely that some form of local civic party will be important: certainly this has been the case in major cities like Montreal, Toronto and Winnipeg.

What does partisanship and intensity mean for the form of local politics? It is much more difficult to generalize on this topic, such are the variations in local arrangements. But in many cases, especially in those western European countries that follow along British lines and run their councils through the extensive use of committees, there is a tendency for the majority party to take all the spoils of victory. In Scandinavia, however, there is a tradition of permitting quite extensive minority party contribution to local policy-making and implementation.

EXPLANATIONS

Understanding these differences in the way in which local government is structured, organized and financed is important if we are to understand the different ways in which the various local government systems reviewed below react to fiscal stress and urban innovation. Explaining these differences is more difficult. It is important to remember just how different in size and population are the various countries we consider. They range from relatively small countries like Denmark and Finland, with a population of around 5 million each, through larger countries like France and Britain, with around 60 or 70 million each, to the USA, with close to 250 million. But compare the USA with Canada, the largest country in area but with a population of no more than 20 million. These differences alone are likely to account for some of the variations just discussed.

But differences in experiences and values – the sum of history – are also likely to produce differences in the way in which local government is organized. For example, the strong non-partisan local tradition in Canada is very much a reaction to what Canadians saw as the worst excesses of partisanship in some US cities at the turn of this century. The strong social democratic traditions of Scandinavia in part help to explain the emphasis in those countries on local government as a service delivery agency, while the strong presence of local politicians at national level in French and Italian politics, which has meant the maintenance of a strong strand of localism in national politics, helps to explain why the structure of local government in those countries has changed so little over time (Tarrow, 1977; Page and Goldsmith, 1987).

The persistence of localism is also a feature of US politics, with the resultant fragmentation of governmental systems that observers of American local politics have come to know and love. But, by contrast, localism – that is, the penetration of national politics by local interests – has largely disappeared from British politics (Bulpitt, 1983). Arguably the same is true for Canada, since, despite the domination of Canada by its large cities in population terms, the local level remains a relatively insignificant one when compared with the importance of the provincial level of government. Again, this difference is likely to result in a different style of local politics, with different emphases and so on.

In effect, what implicitly underlies these points is the need for some understanding of cultural differences between nations which help us to understand the value placed by nations and citizens on the institution of local government and the way in which it operates. The data reviewed here, along with other work (Page and Goldsmith, 1987; Kjellberg and Dente, 1988), suggest that there are a number of different models of local government which can be applied to the countries included in this study. First, there is what might be called the Scandinavian or northern European model, covering Denmark, Finland, Norway and Sweden, and also possibly the UK. In all these countries we find a wide-ranging local government system, based – at least until recently – on the idea of using local government as a service-providing agency for the delivery of welfare state services, themselves the product of a broadly based political consensus across party lines. Second, there is the southern European model, best exemplified here by France and Italy, in which the scope of local government is still relatively narrow, the benefits of the system are delivered largely on a personal and patronage basis, and local politics and local interests are still a pre-eminent feature of national politics. Last, there is the North American model, where the scope of local government remains narrow, where a principal characteristic of the system is its fragmentation, and where benefits are still largely more likely to be delivered to the interest of capital than to individuals. Although the USA and Canada are both examples of these federally based systems, it is important to remember that significant cultural differences between the two countries remain. This is one reason why it is not appropriate to develop these ideas more fully here, but no doubt an explanation based on cultural differences and the kind of models outlined above will help readers to understand the varying ways in which our countries have reacted to different levels of fiscal stress and urban innovation.

REFERENCES

Bulpitt, J. 1983. *Territory and Power in the United Kingdom*, Manchester: Manchester University Press.

Clark, Terry N. and Lorna C. Ferguson 1983. *City Money*, New York: Columbia University Press.

Goldsmith, Michael and Kenneth Newton 1986. 'Local Government Abroad', Report of the Committee of Inquiry into the Conduct of Local Authority Business, *Research*, vol. 4, London: HMSO, pp. 132–158.

HMSO 1986. *Paying for Local Government*, Cmnd. 9714, London.

Kjellberg, F. and B. Dente (eds) 1988. *The Dynamics of Institutional Change*, London: Sage.

Masson, J. and J. Anderson 1972. *Emerging Party Politics in Urban Canada*, Toronto: McClelland and Stewart.

Magnusson, W. and A. Sancton 1983. *City Politics in Canada*, Toronto: University of Toronto Press.

Mouritzen, Poul Erik and Kurt Houlberg Nielsen 1988. *Handbook of Comparative Urban Fiscal Data*, Odense: Danish Data Archives, Odense University.

Page, Edward and Michael Goldsmith (eds) 1987. *Central and Local Government Relations*, London: Sage.

Rhodes, R. A. W. 1988. *Beyond Westminster and Whitehall*, London: Unwin Hyman.

Sanantonio, E. 1987. 'Italy', in Edward Page and Michael Goldsmith (eds), *Central and Local Government Relations*, London: Sage, pp. 107–129.

Tarrow, S. 1977. *Between Centre and Periphery*, New Haven: Yale University Press.

PART II
THE FISCAL CRISIS AND ITS CAUSES

3
What is a Fiscal Crisis?

Poul Erik Mouritzen

In 1985 the city of Copenhagen was under pressure. The budget was running in the red. Without serious changes, the coming years would lead to ruin. City officials led by the Social Democratic mayor launched a campaign with the aim of attracting more money from the central government, a strategy that had proven successful in the late 1970s. However, the government had changed. What had been a successful strategy under the Social Democratic government in 1979 met strong resistance from the conservative coalition government in 1985. After a prolonged public debate and a thorough evaluation of city finances by an independent research institute, the city eventually received temporary relief in the form of an annual special block grant which was to be abolished after five years.

The debate between the Social Democratic city and the conservative minister of the interior exhibited a clash between two different conceptions of a fiscal crisis. For the city, the crisis was obviously a result of uncontrollable forces: increasing demands for municipal services and a decreasing resource base. For the minister, the answer was equally obvious: the city was to blame because it did not adapt to changes in the ability of the private sector economy to sustain a large public sector. The negotiated solution reflected both views. The city of Copenhagen got more money in recognition of the fact that it had special problems. But a condition was attached: the city agreed to reduce expenditures by 2 percent annually over the five-year period, thereby admitting that it could in fact adapt better to changes in the private sector economy.[1]

The same clashes have taken place in several other cities – Liverpool and New York City, just to mention two well-known examples. Also, the two conceptions of a fiscal crisis are often found in the scientific literature. They will be discussed below under the terms 'socio-economic imbalance' and 'maladaption'.

The major purpose of this chapter is to arrive at an operational measure of fiscal crisis which can be applied in cross-national research. Before we go into detail with the exact definition, however, we shall discuss the more exact nature of a fiscal crisis. It is argued that the phenomena of stress and slack possess objective as well as subjective components. The notion of fiscal crisis is deeply rooted in the political process and performs an important function for certain actors in the political system. Also, the perception of crisis is strongly related to expectations and to norms attached to the various roles of the political system.

The Copenhagen case is clearly representative in this sense. For city officials it was important to prove that the city was under pressure, and particularly to stress the socio-economic imbalance perspective which more or less acquits local political leaders of any responsibility. They are subjected to forces of the private market economy over which they have little influence, while central policy-makers are given a major responsibility as it is their task to decide on central government policies which moderate the effects of the market forces. For a conservative government which saw the major problem of the national economy as stemming from the inability of the public sector to adapt to the private market, it was similarly important to stress the maladaption perspective. This perspective is directly linked to local government policies and therefore places the burden of blame for mishaps on the local political leaders. It is their responsibility to adapt to external processes; if they do not adapt, the result is fiscal crisis. Their role is to maintain a financially healthy city and to use the appropriate financial options.

The second part of the chapter deals with problems of operationalization. We consider how previous studies – which in most cases deal with local governments within one country – have defined fiscal stress, and how appropriate these definitions are for comparative analysis. Finally, a definition of fiscal stress as a dynamic phenomenon is formulated and an exact operational measure is arrived at. This definition, which is used extensively throughout the book, incorporates four factors that affect a local government's ability to provide the same level of service over a period of years: the rate of inflation, changes in local needs, changes in the local tax base, and changes in grants from upper-level governments.

FISCAL STRESS AND FISCAL SLACK

Political and financial capital
The basic dilemma of fiscal policy-makers is how to make a trade-off between two types of valuable capital, political and financial. By political

capital we mean the degree to which political leaders are supported by various important groups in society, including voters, administrators and other public employees. By financial capital we mean resources that are potentially free for future public purposes.

A large stock of political capital is indicated when a political leader enjoys the support of all relevant groups in society, when he seldom meets resistance as a consequence of his policies and, ultimately, when his seat in the elected body is uncontested during elections.

A large stock of financial capital is indicated by a low degree of exploitation of the taxation base (a low rate of taxation), low debt and a large stock of liquid capital.

Political and financial capital can be exchanged. The political capital can be increased via the distribution of material benefits, that is, via the use of financial capital. The financial capital may, similarly, be increased at the expense of the stock of political capital. This is often what happens when cities reduce their expenditures, for instance in order to build up liquid assets for future use.

It is immediately clear that political support is not simply rooted in the distribution of visible benefits to members of society. There is a basic assumption that political leaders are able to manipulate or influence specific support through the distribution of material benefits. However, it must also be recognized that they often have other means at their disposal which do not rely solely on the availability of financial resources. Regulative or symbolic policies can often be pursued at no cost but with formidable effects on the stock of political capital.

In the late 1970s the Social Democratic mayor of the municipality of Vissenbjerg (my home town) made himself a nationally known politician by resisting pressures from certain central government agencies which wanted all municipalities to report the social security numbers of people on welfare. By focusing the nation's attention on this little town, his popularity locally was at a peak before the election in 1981. He then engaged in what, judged by all political science textbooks, would be a suicidal policy. Two months before the election, the Social Democratic majority decided to increase the tax rate by almost 3 percentage points (from about 18 to 21 percent), making the rate of taxation in the municipality among the highest in the country. The Social Democrats lost one seat and the majority. The Mayor then engaged in the next – again judged by the textbook – self-defeating step. He entered into an agreement with a council member from the anti-tax Glistrup Progress Party, who agreed to vote for the Social Democratic mayor in return for the chairmanship of the (for Social Democrats) traditionally important social welfare committee! The mayor was re-elected in 1985, but in 1989 his stock of political capital had evidently depreciated. Despite promises to the contrary made during the 1985 election, he had tried to close down a couple of small village schools. His general popularity was not able to compensate for broken promises, and the mayorship went to the opposition.

The example shows how, as a result of a very large stock of political capital built up through symbolic behavior, a political leader may maneuver through a fiscally hard period pursuing policies that are, by most standards, deemed to be self-defeating. But it also shows that voters have a short-term perspective: political capital is not everlasting. Realizing the existence of these additional mechanisms does not make studies of fiscal policy-making irrelevant. However, the fact that politicians may deliberately use regulative and symbolic policies to minimize loss of specific support in fiscally hard times points towards some of the limits of a partial analysis. Often we find cities that respond in a completely unpredictable way. Such deviations from the typical pattern can often be explained by certain peculiarities such as, for example, a very popular (or, for that matter, unpopular) mayor.

The idea of a stock of political and financial capital may be represented in the form of a diagram (cf. Figure 3.1). Any political leader or local

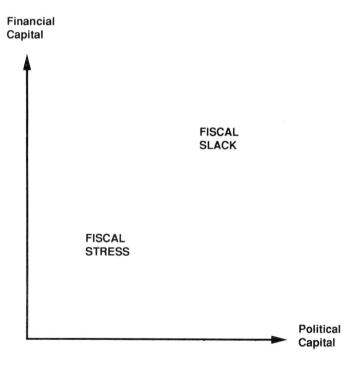

Figure 3.1 *Fiscal stress and slack as a function of political and financial capital*

government would be positioned somewhere in this diagram. When we talk about fiscal stress or fiscal slack as a description of the situation at a certain point in time, we are thinking along these lines. Fiscal slack implies a large stock of both political and financial resources; you are in the north-eastern quadrant of the diagram. Fiscal stress implies few political and financial resources; you are in the south-western quadrant of the diagram.

The position in the diagram of a politician or a local government may change. This can happen in three ways. During a period of stable external conditions, a political leader may choose to build up political capital at the expense of financial capital. This is typically what happens before an election. This kind of change is indicated by a movement along the line *ab* in Figure 3.2. This line is similar to the budget line in consumer theory.

Also, a political leader may be able to increase his political capital without utilizing his stock of financial capital. Through symbolic behavior or regulative policies, he may be able to 'buy' additional support with no consequences for his future financial resources. This kind of change is indicated by a shift in the slope of the budget line (from *ab* to *ac*). Because the price (in terms of exploitation of the resource base) of a given amount

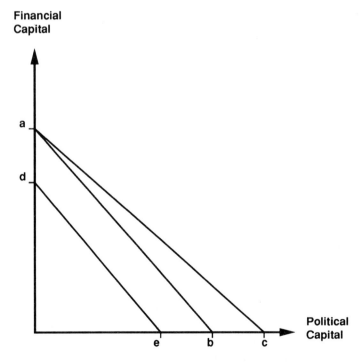

Figure 3.2   *Changes in the stock of political and financial capital*

of political support has decreased, this change is similar to the price effect in consumer theory.

Finally, changes in the external conditions may force upon political leaders changes in their stock of political and financial capital. This may happen if, for instance, central government reduces (or increases) its support to local governments. Such changes are parallel to the income effect in consumer theory. Decreasing grants thus imply a change in the budget line from *ab* to *de*, implying that the net stock of political and financial capital has been reduced. When we talk about fiscal stress as a change phenomenon, we are thinking about such changes in the budget line. When, for instance, the position of the budget line changes from *ab* to *de*, it is indicative of increasing resource scarcity. Fiscal stress implies that the fiscal constraint has tightened; that is, the budget line moves towards the lower left corner of the diagram. Fiscal slack implies a shift in the opposite direction. Political and financial capital can be increased at the same time.

The politics of fiscal crises

In a report from 1980, the Danish Association of Local Authorities discussed the emerging 'fiscal squeeze' and what to do about it (Kommunernes Landsforening, 1980). Looking back, it did not make much sense to talk about a fiscal squeeze in the midst of a period (1978–80) where current expenditure was increased by almost 13 percent in real terms and where available resources with constant rate of taxation increased by almost 5 percent. By any objective standard, there was no fiscal crisis.

This example, however, merely underscores the fact that fiscal crises, besides being phenomena rooted in the objective fiscal reality of local governments, also are symbolic phenomena with very important political functions for local leaders.[2] The talk about a fiscal squeeze in Denmark in 1979, therefore, can be understood only in a specific political context. This was the first year that central government through a new budgeting system was able to get a detailed insight into local finances (including long-term budgets), a fact that totally changed the rules of the game between central and local governments. The report and the following debate about the fiscal squeeze can thus be perceived as one among several strategic moves made by the Association of Local Authorities to improve its bargaining position vis-a-vis central government.

Most central governments do exercise a considerable degree of control over local finances, for instance via the amount of grants floating to local governments, limits on taxation and borrowing, or limits on spending. As the example above shows, this gives local governments a strong incentive always to paint a negative picture of their fiscal condition, often a picture that has no relation at all to objective conditions.

Norway represents an excellent example. During the first part of the 1980s, Norwegian local governments were probably among the luckiest in the world (cf. Chapter 4). However, all evidence suggests that they were complaining no less than local governments in other countries. Most people in Norway, including policy-makers and scholars, did believe that local governments had been under heavy pressure. When in the fall of 1987 the first comparative results were published by the Fiscal Austerity and Urban Innovation project, they created a heated national debate. Eventually, this led to a reduction in grants to local governments.

Holland is also interesting. The FAUI survey was paid for by the Dutch Association of Local Authorities. After the first analyses the data files were classified and no further analyses were made public.[3] The reason: the survey painted a much brighter picture of the conditions of local governments than the one maintained by the Association. Obviously, the association did not want to weaken its future bargaining position vis-a-vis central government by having irresponsible academics broadcasting this information around the world.

Also, individual local governments have an interest in fertilizing the notion of crisis. Often they engage in negotiations with state agencies for more resources. In recent years, with the imposition of revenue or spending limits in many countries, many have been seeking to negotiate exemptions from the general regulations. Such exemptions are obviously much easier to obtain if a local government is under fiscal pressure.

More importantly, the notion of crisis plays an important role in the local political process. If the local constituency can be convinced that there is in fact a crisis (and, even better, if it can be convinced that central government is to blame), the political costs of many unpopular actions taken by local officials can perhaps be reduced. Thus, much talk about fiscal crises can be perceived as symbolic behavior that local political leaders engage in with the purpose of reducing political losses. In other words, to overemphasize the extent of the fiscal crisis through symbolic behavior is one of the ways in which political leaders may in fact reduce some of the unfortunate consequences of the actual/objective crisis in terms of loss of political capital.

This discussion does not imply that there are never real-scale crises. There are. It should merely make us cautious about the kind of observations we make in order to single out the extent of fiscal stress or slack.

The psychology of fiscal crises
To the extent that 'fiscal crises' perform these politically important functions, local leaders will most often realize that the crises are really non-existent. It is, however, equally true that some 'crises' do in fact exist

as real phenomena in the minds of local actors. This is because crises are deeply rooted in the expectations of people.

According to Rose (1980: 211), 'frustration is the subjective consequence of a "squeeze" between what is expected and what is received'. Such a squeeze can (still according to Rose) arise in three different ways besides objective deprivation: expenditures may grow less than historic growth; expenditures may grow less than announced (in multiyear budgets); or cash limits may force agencies to spend less than originally appropriated.

From 1978 to 1979, local governments in Denmark experienced a record expenditure growth of 7–8 percent (in real terms). When limits on spending growth of 3 percent were agreed upon between central government and the Association of Local Authorities, this created a rather massive feeling of crisis in many municipalities (some of which had been able to increase spending by more than 10 percent from 1978 to 1979 without even raising tax rates). In subsequent years the limits were reduced, and from 1981 to 1982 the limit aimed at 1 percent real growth. Using this method of gradually reducing the expectations of local governments and recipients of local services, the Association and the government were able to slowly reduce the feeling of crisis. It is symptomatic that in 1982 there was not much talk of a fiscal squeeze.

A feeling of crisis results from relative deprivation. Expectations as they are formed by previous experience or by promises made in multiyear budgets may not be met. There is, however, another source of relative deprivation which can also account for certain parts of the fiscal crisis. This is when other actors are doing better than you are doing, 'other actors' being other local governments or other actors within the same local government. As there are always bound to be losers in fiscal policy-making (absolutely or relatively speaking), there will always be a tendency for an asymmetric perception of the fiscal well-being of government to develop.

Changes in the locational patterns of private businesses may, for instance, produce large regional inequalities among local governments. Localities that are better off are generally not likely to advertise their luck (for reasons discussed previously), while localities that are worse off (whether absolutely or relatively) will do everything to make public their bad luck. In other words, even though, at the aggregate level, local governments in a country are not fiscally squeezed, a combination of politics, psychology and real-scale fiscal crises in certain localities is likely to produce a pervasive feeling of crisis.

Organizational roles and fiscal crisis
Within the individual local government, the same kind of process is likely to take place. Some sectors, departments or bureaus will always lose,

whether relatively or absolutely. So even if aggregate expenditures are kept constant, some actors will always have a reason to talk about the fiscal squeeze because they have become relatively deprived.

There is, however, an additional reason why at least some actors within a local political system will always perceive some sort of crisis. This is due to the fact that the definition of 'crisis' is a product of roles and policy responses. Actors in a local government are likely to define crisis differently depending on the extent to which the policy actions taken by the city council corresponds to their own preferences. A socialist political leader will perhaps define 'crisis' as a situation where expenditures are unable to meet rising demands (crisis is due to an expectation gap; cf. below), while a conservative political leader will see a 'crisis' if expenditure grows faster than the private sector economy can support (crisis is due to maladaptation; also cf. later).

Results from the Danish Fiscal Austerity Project support this perspective. There was no significant correlation between the objective measure of fiscal stress (as defined later) and the subjective evaluations made by the finance directors of the municipality ($r = -0.01$!). However, the subjective feeling of crisis was explained to a large extent ($R^2 = 0.91$) by eight response variables, and the causal relationship was in the direction expected, considering the role of a finance director: the more the municipality had used percentage across-the-board cuts, had cut expenditures supported by conditional grants from central government and had increased loans, the greater his tendency to see a crisis. These three strategies probably do not make a lot of sense to a finance director whose responsibility it is to make ends meet in the short as well as the long run and to insure that resources are used efficiently. On the other hand, the finance director tended to underplay the extent of the crisis, the more the municipality had reduced personnel, had established joint purchasing agreements and had omitted regulation for price/wage increases in certain areas.[4]

MEASURING FISCAL SLACK AND STRESS

The previous discussion indicates how difficult it is to give the fiscal crisis objective substance. Objective financial conditions mix with the fiscal, regulative and symbolic policies, political tactics and perceptions of various actors. In order to be able to approach an operational definition, let us first consider some definitions found in previous research and set up a few explicit criteria.

Previous research

Most studies of fiscal stress share a common concern in the sense that they deal with imbalances. However, the focus may be different. Here we evaluate three concepts of fiscal stress often found in the literature. They differ fundamentally with respect to how they define the imbalance. In order to pinpoint these differences, we borrow from systems theory three analytical categories: demands (needs or wants), resources, and fiscal policy outputs (cf. Mouritzen, 1985: 13f.).

One tradition of research conceives of fiscal stress as a socio-economic imbalance. There is a mismatch between what citizens want in relation to what the private sector economy can support. In other words, fiscal stress is a situation where there is an imbalance between needs and resources. This concept is found in studies focusing on short-term changes in resource levels (Levine et al., 1981: 43ff.; Schick, 1980: 114ff.) or tax bases and intergovernmental grants (Wolman and Davis, 1980: 2f.), as well as long-term changes in social and economic conditions as indicated in the 'migration and tax-base erosion model' (Rubin, 1982). Structural differences in social and economic conditions have also been investigated (Nathan and Adams, 1976) and are usually the focus of governmental research which aims at developing indicators of needs used in the distribution of grants to local governments (Cameron and Lotz, 1981; Kamer, 1983: 4–5; Aronson, 1984).

It is important to emphasize that this perspective does not say anything about the financial situation of a city. It does not define or determine the 'fiscal health' of a local government (Clark, 1977: 54). The Advisory Commission for Intergovernmental Relations has quite appropriately applied the term 'community distress' for indicators that measure 'the relative economic and social condition of people, the places they live and businesses' (ACIR, 1985: 2). The strategy of the city of Copenhagen (cf. the introduction to this chapter) was to emphasize the socio-economic imbalance perspective.

A considerable body of literature – probably the largest part – directly links the input side of the local political system to the output side. Fiscal stress is conceived of as an imbalance between fiscal policies and available resources. Stress is the result of maladaption. This is the perspective held by the Ministry of the Interior in the debate with the city of Copenhagen about economic support.

The literature contains a long list of indicators which (implicitly or explicitly) are based on this maladaption perspective. A number of studies discuss short-term measures of deficits which may be indicative of a situation where expenditures increase faster than resources (ACIR, 1985: 12; Rubin, 1982; Chan and Clark, 1982: 7; Clark, 1977: 55; and for a critical assessment Wolman and Davis, 1980: 3). More dominant, however, are ratio measures in which expenditures, debt and liquid assets are related to resource indicators like own-source revenues, tax base and population (Clark and Ferguson, 1983: 52 ff.; Chan and Clark, 1982: 9 ff.; Kamer, 1983: chs. 4 and 5; Kelso and Magiotto, 1981; Hunter, 1982:

143; Howell and Stamm, 1979; Kennedy, 1984: 97; Newton, 1980: 12). Such ratio measures usually involve level as well as change measures (cf. Clark and Ferguson, 1983: 45–46).

A few studies apply a third concept of fiscal stress. We call it 'the expectation gap approach' because it focuses on the imbalance between demands, needs or wants on the one side and fiscal policies on the other. ACIR (1985: 2) uses the term 'citizen fiscal distress' to denote a situation where 'a local government is unable to provide a basic package of services and to collect the revenues necessary to pay for them'. Kamer (1983: 72) uses the conventional term of 'tax efficiency', assuming that higher degrees of efficiency lead to higher levels of citizens' satisfaction.

Almost all studies of financial problems in local government use objective measures of fiscal stress. In only a few instances have studies been based on subjective indicators, such as the evaluations of central actors of the seriousness of the fiscal crisis in a city (Skovsgaard, 1985; Nathan et al., 1975).

Most studies of urban fiscal stress seem to involve change measures, where changes are typically measured over a short period, such as three to four years. Less often we find level measures, which seek to rank cities at one particular point in time according to some criteria, for example the level of taxation, debt per capita or expenditure divided by revenue. The two kinds of measures are likely to produce quite different results (Mouritzen, 1989).

Looking for a definition: some criteria of choice
In the present context, the definition and its operationalization must meet three criteria:

1 We need a measure that can be applied to aggregate comparative analyses as well as city-level comparative analyses. In other words, the indicator should be able to measure – on an interval scale – fiscal stress and slack in local governments as a whole in various countries and in individual local governments within various countries. In the first case the unit of analysis is the country, while in the latter it is the individual locality.
2 The fiscal stress/slack indicator should meet the requirements implied by the theoretical discussion above; that is, it should ideally reflect the stock of and changes in financial and political capital.
3 We are interested in the ways in which local governments respond to fiscal stress and slack. For that reason, the indicator should be independent of policies pursued by local governments.

These requirements are quite demanding, considering the particularities of the local government systems in the 10 selected countries. Each local governmental system is unique with respect to the distribution of responsibilities between layers of government, degree of autonomy, grant system, system of taxation and degree of consolidation. Precise operational measures of theoretical concepts are almost impossible to obtain without adapting them to the particular setting in each country.

One possible solution to this problem would be to abstain from objective measures and rely on the evaluations of central actors which are available for most countries through survey research. For reasons discussed earlier, this procedure is not appropriate.[5] Since there is a tendency to perceive a fiscal crisis anywhere, subjective perceptions would not really grasp the major differences between countries. As the information would in most cases come from one type of actor, it would furthermore tend to be partially determined by the policies pursued. A subjective measure of fiscal stress would therefore violate criteria 1 and 3 above.

The distinction between level and change measures was mentioned above. By a 'level measure' we mean an indicator that tells us something about the fiscal conditions of a local government at one particular point in time (cf. Figure 3.1), for example debt per capita in 1982, or expenditures divided by city wealth in 1986. A 'change measure', on the other hand, gives us an indication of how the fiscal conditions have changed over a period of years; an example could be percentage change in expenditures from 1982 to 1986 divided by percentage change in city wealth over the same period. Which one is most applicable in comparative research?

While level measures may have a role to play in within-country studies, they are very difficult to apply in comparative research. The only example found in the literature is Wildavsky's classification of contexts for budgetary conflicts (Wildavsky, 1975: ch. 12), where he identifies the position of various countries according to two criteria: the containment-of-conflict ratio and the support-on-spending ratio. The purpose is to identify the political surplus (or deficit) and the economic surplus (or deficit) in a country, very much corresponding to the ideas proposed above about a stock of political and financial capital. This procedure, ambitious and impressive as it may be, is not feasible in the present context, where local governments are the object of analysis and where we do not have enough reliable information to identify the position of local governmental systems in the various countries along the two dimensions.[6]

In comparative within-country studies it is often possible to take a short-cut via fiscal and socio-economic data which allows identification of the position of individual local governments along the two dimensions of political and financial resources. In comparative between-countries urban studies this is probably not possible, because of the different functional responsibilities of local governments. The fact that local governments, in terms of scope and range of functions, are very different across countries makes it almost impossible to construct sensible level measures of fiscal stress.[7]

To sum up, level measures of fiscal stress were not found applicable because they would violate criterion 1. As a consequence, we will have to compromise on criterion 2, because we are unable to measure the stock of financial and political resources of local governments.

Above we considered three perspectives of fiscal stress often found in the literature: the socio-economic imbalance perspective, the maladaption perspective, and the expectation gap perspective. The maladaption approach clearly grasps only one dimension. Fiscal stress, according to this approach, 'involves a reduction in the number of financial options available to a city' (Rubin, 1982: 2); that is, it stresses the stock of financial capital while the political capital is left out. Maladaption does not reflect 'dissatisfaction by citizens whose services are trimmed' (Clark and Ferguson, 1983).

In contrast, the expectation gap approach merely deals with political resources while the financial capital is ignored. It focuses on citizen satisfaction with urban services. Whenever a gap develops between what citizens expect and what the government provides, the result is a decrease in the stock of authoritative resources.

In essence, for both the maladaption and the expectation gap approach, fiscal stress is a product of policies pursued by governments. Both approaches relate 'what government does' to what the economy can support or what citizens want, respectively. In contrast, the defining factors in the socio-economic imbalance perspective are external to policy-makers. Local political leaders cannot – at least, not in the short term – manipulate either the resource base or the needs and demands of citizens. Socio-economic imbalance is a causal factor while maladaption and expectation gaps are effects. Concepts of fiscal stress which have a built-in policy element are methodologically problematic because conclusions about the responses to fiscal stress may be true by definition.[8]

To sum up, we choose a socio-economic imbalance concept of fiscal stress arising from criteria 2 and 3.

A definition and its operationalization
According to our criteria, we aim at a concept of fiscal stress that can be measured objectively in several countries, is based on a socio-economic imbalance perspective, and measures changes over a period.

This is exactly what the definition proposed by Wolman and Davis does. Fiscal stress is

> a situation in which a local government, faced with the necessity of achieving

a balance between revenues and expenditures, must in time choose either to (1) increase taxes through changes in the tax rate or structure in order to maintain existing real expenditures and service levels, (2) reduce real expenditures from the level of the previous year or (3) engage in some combination of these activities. (Wolman and Davis, 1980: 1)

This definition corresponds to the more abstract definition introduced earlier, where fiscal stress was conceived of as a situation where politicians faced a painful choice: they had to reduce their stock of either financial or political capital.

The definition is parallel to the concept proposed by Levine (1980: 4) and corresponds to the thinking of Peters and Rose (1980), who distinguished between three situations which a government can face: a fiscal dividend, no dividend, and overloaded government. A fiscal dividend thus implies that services can be expanded with no costs for the taxpayers, while overload constitutes a situation like the one described in the definition above.

In order to measure fiscal stress and its opposite, fiscal slack, we will incorporate four factors that affect a government's ability to provide the same level of service in the current year (t) as in a previous year ($t-x$).[9] First, inflation increases the cost to the government of paying for the services it provides. Thus, governmental revenues must increase by at least the same rate as inflation. Next, expenditure needs may change. At the simplest level, population might increase and additional revenues would be required in order to provide services to the increased population. Municipal government revenue sources may also change: government grants may increase or decrease or the local tax base may expand or contract, resulting in increases or decreases in local tax revenues derived from taxes levied at the same rate as in the previous year.

These four factors are used to create an index measuring the degree of fiscal slack a municipal government faces. (We call it a measure of slack because an increasing value is indicative of an improvement in local finances.) If municipal revenue (government grant plus local tax revenues raised by levying the same tax rate as previously) increases sufficiently to cover the additional revenue required as a consequence of inflation and changes in need, then the municipal government will be in a slack situation: it will be able to provide the same (or higher) levels of service per capita to its citizens without raising its tax rate. If, however, municipal revenues do not increase sufficiently to cover the additional revenue needs (or if they decline), the municipal government is in a situation of fiscal stress: it must either raise the tax rate in order to continue to provide the same level of services per capita as before, or cut expenditures, thereby reducing service levels (unless productivity increases sufficiently to compensate fully for the expenditure reduction).

The definition of slack is operationalized in the following formula:

$$SLACK = [(PRREV_t / REV_{t-x}) / CN] \times 100,$$

where

$$PRREV_t = (T_{t-x} \times B_t) + G_t$$
$$REV_{t-x} = [(T_{t-x} \times B_{t-x}) + G_{t-x}] \times r$$

CN = change in expenditure needs from $t–x$ to t
T = tax effort
B = tax base
G = total grants
r = (1 + rate of inflation).

The *SLACK* indicator measures the extent to which revenues with constant tax effort are able to keep pace with changes in expenditure needs and the rate of inflation. A score above 100 indicates a situation where a local government can increase service levels (expenditures controlled for changes in needs and rate of inflation) with a constant rate of taxation over the period $t–x$ to t. A score below 100 shows that the government, if it chooses to operate at a fixed level of taxation, is forced to cut back real service levels. A score of 90, for example, means that a local government has lost 10 percent of the revenues necessary to maintain the service and taxation level from $t–x$ to t.

When, in the following chapters, fiscal stress is mentioned, it is to indicate scores below 100 on the *SLACK* index. Fiscal slack, on the other hand, is indicative of a score above 100 on the *SLACK* index; cf. Figure 3.3. The period $t–x$ to t can be of any length from 1 year to 4, 10 or 20 years.

Change in expenditure needs is very difficult to measure precisely in aggregate comparative analyses. It was decided to use the crude measure of population change in the national-level analyses. Thus, a 5 percent increase in population over the period would result in a *CN* value of 1.05. In the city-level analyses an index was constructed which is based on change in dependent population, that is, the age categories towards which the major local services are addressed, typically population below 17 years and above 65.[10]

100

Fiscal Stress   Fiscal Slack

Figure 3.3   *The fiscal slack indicator*

The rate of inflation measure is for most countries based on the government final consumption index. In a few countries an index was used which specifically shows wage and price increases in local government expenditure.

We use actual figures for the city of Copenhagen in order to illustrate the computations involved. From 1982 to 1986 expenditure needs were reduced by 4.1 percent while the cost index rose by 22 percent. General grants were reduced from DKr 2,805 to DKr 2,236, while the local tax base increased from DKr 24,037 to DKr 28,776 (all figures in millions). Tax effort in 1982 was at 19.67 percent.

First we calculate projected revenue in 1986, which shows the amount of resources available to the city council of Copenhagen had it applied the same tax rate as in 1982:

$$PRREV_{86} = (T_{82} \times B_{86}) + G_{86}$$
$$= (0.1967 \times 28776) + 2236 = 7896.$$

Next we find how much revenue is necessary in order keep pace with inflation

$$REV_{82} = [(T_{82} \times B_{82}) + G_{82}] \times r$$
$$= [(0.1967 \times 24037) + 2805] \times 1.22 = 9190.$$

Finally, we calculate the *SLACK* score as

$$SLACK = [(PRREV_{86} / REV_{82}) / CN] \times 100$$
$$= [(7896 / 9190) / 0.959] \times 100 = 89.59$$

where 0.959 reflects a reduction in expenditure needs over the period of 4.1 percent.

Relying on this measure, we would maintain that the city of Copenhagen was in fact exposed to forces outside its control. In other words, the fiscal crisis perspective used by the city in the 1985 negotiations with the government was grounded in reality.

In order to be able to obtain data from many countries, some fiscal components had to be left out. It could be argued that fees and charges – which in most countries add up to between 15 and 25 percent of total revenues – should be included. In this case fees and charges would have to be considered as a tax under the assumption that a real increase was equivalent to an increase in tax effort. It was decided to exclude fees and charges from the formula because they are levied on quite different objects in the various countries, ranging from day-care and housing to utilities, sewage, etc. The tax effort, therefore, is defined as total taxes divided by tax base.[11]

Also, the formula does not consider the extent of deficit financing (or the opposite) in the base year $(t–x)$. At the aggregate level this probably does not affect the results seriously. At the city level of analysis it may produce certain errors, because a local government that draws heavily on its liquid assets in the starting year may have operated at an unnaturally low level of taxation. Finally, it was not possible to correct for possible changes in the distribution of tasks between levels of government. This may result in certain flaws particularly at the aggregate level of analysis, flaws that have to be taken into consideration when the data are interpreted.

Despite these problems, we maintain that the *SLACK* indicator is a fairly precise measure of the extent to which localities can maintain current services without having to raise taxes, or, more generally, of the extent to which political leaders are forced to deplete their stock of financial and political resources.

NOTES

1 Incidentally, exactly the same debate took place in the summer of 1990. The city did not reduce spending at all (for instance, the number of employees increased over the 1985–90 period by 2–3 percent!) and was, as a consequence, in an extremely bad situation in 1990 because the special grant was being abolished. This time the Minister firmly opposed new special grants to the city.

2 Also, for academics the local fiscal crisis may have important functions to fulfill. When the Danish Fiscal Austerity Project started in 1980, the various funding sources, of which the Local Government VAT Fund was the most important, were approached with an explicit reference to the 'Fiscal Squeeze'!

3 The Dutch survey data reported in Mouritzen and Nielsen (1988) were actually calculated from the one article published from the Dutch project, as there was no way we could get access to the frequencies.

4 The question used to tap the subjective feeling of crisis asked about the general feeling of 'size and seriousness of fiscal problems', allowing the finance director to choose among four categories: (1) None or few problems; (2) Some but not very big problems; (3) Big but only short-term problems; (4) Big, long-term problems. This indicator was treated as an interval scale variable in a standard regression. Two additional variables – use of liquid assets and cuts in maintenance – were not significantly related to the perception of fiscal crisis.

5 Subjective indicators may, of course, be relevant for certain purposes, but in most cases they are bound to overemphasize the extent of the fiscal crisis. Nathan et al.'s study deliberately asked the respondent about the financial status of his local government compared 'with that of nearby units or others of similar type within the state'. To the extent that the respondents are able to correctly judge the relative condition of their own jurisdiction, this will lead to a 50–50 division into fiscally troubled and fiscally non-troubled

local governments no matter how the developments are in the total local government sector as such. In comparative terms this might well produce the same 'score' in a (by objective standards) fiscally fortunate Norwegian municipality and a fiscally pressed English local government; cf. the *SLACK* scores for the UK and Norway in Chapter 4.

6 Probably the only way to replicate the Wildavsky procedure is to conduct extensive citizen surveys in several countries. Another piece of research by Richard Balme (cf. Chapter 7 below), where he reports on the preferences of political leaders and citizens, is a first step in this direction.

7 The so-called functional performance approach (cf. Clark et al., 1982) represents a distinct methodology to overcome the difficulties arising from differences in functional responsibilities. While this method seems to produce sensible results in one country (the USA), it is not immediately transferable to cross-national analysis. The weighting procedure, for instance, requires the same monetary standard and is therefore sensitive to changes in currency rates.

8 This point is discussed in detail in Mouritzen (1989).

9 The exact operational definition was developed with the assistance of Hal Wolman.

10 This is true for the four Nordic countries. For France and the USA we used change in overall population because of the unavailability of age-specific data.

11 In most countries tax base was defined as personal income, under the assumption that most taxes (whether income, property or taxes on goods) are paid out of people's incomes.

REFERENCES

ACIR 1985. *The States and Distressed Communities: The Final Report*, Washington: Advisory Commision of Intergovernmental Relations.

Aronson, J. Richard 1984. 'Municipal Fiscal Indicators', in James H. Carr (ed.), *Crisis and Constraints in Municipal Finance*, New Brunswick, NJ: Rutgers University Press.

Cameron, Gordon and Jørgen Lotz 1981. *Measuring Local Government Expenditure Needs: The Copenhagen Workshop*, Paris: OECD.

Chan, James L. and Terry N. Clark 1982. *Measuring Municipal Fiscal Strain*, Urbana, Ill.: University of Illinois and University of Chicago.

Clark, Terry N. 1977. 'Fiscal Management of American Cities: Funds Flow Indicators', *Journal of Accounting Research*, 15: 54–94.

Clark, Terry N. and Lorna C. Ferguson 1983. *City Money: Political Processes, Fiscal Strain, and Retrenchment*, New York: Columbia University Press.

Clark, Terry N., Lorna C. Ferguson and Robert Y. Shapiro 1982. 'Functional Performance Analysis: A New Approach to the Study of Municipal Expenditures and Debt', *Political Methodology*, 8(2): 87–123.

Howell, James M. and Charles F. Stamm 1979. *Urban Fiscal Stress: A Comparative Analysis of 66 US Cities*, Lexington, Mass: Lexington Books.

Hunter, William J. 1982. 'The Impact of Labor Costs on Municipal Finances', *Public Choice*, 38: 139–147.

Kamer, Pearl M. 1983. *Crisis in Urban Public Finance: A Case Study of Thirty-Eight Cities*, New York: Praeger.

Kelso, W.A. and M.A. Magiotto 1981. 'Multiple Indicators of Financial Instability in Local Governments', *International Journal of Public Administration*, 3(2): 189–218.

Kennedy, Michael D. 1984. 'The Fiscal Crisis of the City', in Michael P. Smith (ed.), *Cities in Transformation: Class, Capital and the State*, Sage Urban Affairs Annual Reviews, vol. 26, Beverly Hills: Sage.

Kommunernes Landsforening 1980. *Udgiftspresset og de økonomiske muligheder i kommunerne*, Copenhagen: Forlaget Kommuneinformation.

Levine, Charles H. (ed.) 1980. *Managing Fiscal Stress: The Crisis in the Public Sector*, Chatham, NJ: Chatham House.

Levine, Charles H., Irene S. Rubin and George G. Wolohojian (eds) 1981. *The Politics of Retrenchment: How Local Governments Manage Fiscal Stress*, London: Sage.

Mouritzen, Poul Erik 1985. 'Concepts and Consequences of Fiscal Strain', in T.N. Clark, G.M. Hellstern and G. Martinotti (eds), *Urban Innovations as Response to Urban Fiscal Strain*, Berlin: Verlag Europaische Perspektiven, pp. 13–33.

Mouritzen, Poul Erik 1989. 'Dimensions of Fiscal Stress', *Research in Urban Policy*, 3: 141–164.

Mouritzen, Poul Erik and Kurt Houlberg Nielsen 1988. *Handbook of Comparative Urban Fiscal Data*, Odense: Danish Data Archives, University of Odense.

Nathan, Richard P. and Charles Adams 1976. 'Understanding Central City Hardship', *Political Science Quarterly*, 91: 47–62.

Nathan, Richard P., Allen D. Manvel and Susannah E. Calkins 1975. *Monitoring Revenue Sharing*, Washington DC: Brookings Institution.

Newton, Kenneth 1980. *Balancing the Books: Financial Problems of Local Government in West Europa*, London: Sage.

Peters, Guy and Richard Rose 1980. 'The Growth of Government and the Political Consequences of Economic Overload', in Charles H. Levine (ed.), *Managing Fiscal Stress: The Crisis in the Public Sector*, Chatham, NJ: Chatham House.

Rose, Richard 1980. 'Misperceiving Public Expenditure: Feelings About "Cuts"', in Charles H. Levine and Irene Rubin (eds), *Fiscal Stress and Public Policy*, Sage Yearbooks in Politics and Public Policy, vol. 9, London: Sage.

Rubin, Irene 1982. *Running in the Red: The Political Dynamics of Urban Fiscal Stress*, New York: State University of New York Press.

Schick, Allen 1980. 'Budgetary Adaptions to Resource Scarcity', in Charles Levine and Irene Rubin (eds), *Fiscal Stress and Public Policy*, Sage Yearbooks in Politics and Public Policy, vol. 9, London: Sage.

Skovsgaard, Carl Johan 1985. 'Budgeting Innovations in Danish Municipalities under Fiscal Austerity', in T. N. Clark, G. M. Hellstern and G. Martinotti (eds), *Urban Innovations as Response to Urban Fiscal Strain*, Berlin: Verlag Europaische Perspektiven, pp. 129–138.

Wildavsky, Aaron 1975. *Budgeting: A Comparative Theory of Budgetary Processes*, Boston: Little Brown.

Wolman, Harold and Barbara Davis 1980. *Local Government Strategies to Cope with Fiscal Stress*, Washington, DC: Urban Institute.

4

Was There a Fiscal Crisis?

Poul Erik Mouritzen and Kurt Houlberg Nielsen

In the late 1970s a group of European scholars surveyed the fiscal conditions of local governments in a number of countries, the so-called 'Volkswagen project' (Newton, 1980). Danish, Swedish, Norwegian, West German, UK and Italian local governments were ranked according to five measures: changes in gross debt, surplus or deficit before loan transactions, debt charges as a percentage of total expenditure, real increases or decreases in expenditure and the growth ratio of taxes and grants to local spending (Newton, 1980: 11).

The financial situation of local governments in the six countries was, in summary, described as:

Denmark	Excellent
Sweden	Very good
Norway	Good
West Germany	Satisfactory
UK	Poor
Italy	Critical

In this chapter we report on a follow-up study seeking to answer questions similar to those posed by the Volkswagen group. In doing so we focus on the objective fiscal conditions of local governments, thus deliberately abstaining from a discussion of the political and psychological aspects of fiscal crises. We use the measure of *fiscal slack* developed in the previous chapter, which incorporates four factors that affect a government's ability to provide the same level of service in the current year (t) as in a previous year ($t-x$): inflation, changes in expenditure needs, tax base, and grants. If municipal revenue (government grant plus local tax revenues raised by levying the same tax rate as previously) increases sufficiently to cover the additional revenue required as a consequence of inflation and changes in needs, then the municipal government will be in a slack situation; i.e., the *SLACK* indicator will take a value above 100. If the *SLACK* score is below 100, municipal revenues do not increase sufficiently to cover the additional revenue needs (or they decline). In this case the municipal government is in a situation of *fiscal stress*, that is, a situation where it must either raise the tax rate in order to continue to provide the same level of services per capita as previously, or cut expenditures.

Based on aggregated national data, the first part of the chapter shows how and when municipalities in the 10 selected countries have been struck by a fiscal crisis over the period 1978–86. In the second part we search for the determinants of fiscal crisis. We try to explore some possible links between the health of the national economy and the financial constraints imposed upon municipalities. This part is also based on aggregated national data. In the final part of the chapter we turn to city-level data in order to show how cities within countries differ with respect to financial conditions.

TRENDS IN LOCAL GOVERNMENT FISCAL CONDITIONS

In Table 4.1 the values on the *SLACK* indicator are shown for the 10 selected countries starting in 1978. The value for each year takes as its starting point (base year) 1978; in other words, the rate of taxation used in the calculations is always from this year.[1] In order to make trends and patterns more visible, the *SLACK* indicator is presented graphically in Figures 4.1 and 4.2. Figure 4.1 covers the five countries with the most favorable development in the fiscal constraint on municipalities (the 'slack' countries). Figure 4.2 covers the five countries that have, over the period or part of the period, experienced a substantial deterioration in the financial conditions of local government (the 'stress' countries).

Table 4.1   *Fiscal slack in municipalities*

	1978	1979	1980	1981	1982	1983	1984	1985	1986
Denmark	100	101.8	104.8	105.5	108.1	106.9	105.7	100.7	100.4
Norway	100	103.6	109.5	110.9	114.9	116.4	123.4	131.3	–
Sweden	100	104.8	103.4	105.5	105.4	105.3	108.1	108.6	111.0
Finland	100	104.7	107.7	109.7	114.4	123.0	131.1	137.6	–
Germany	100	101.8	103.1	99.6	97.8	96.2	99.2	103.0	106.1
UK	100	101.9	97.2	89.3	86.7	90.8	91.0	92.7	–
France	100	104.4	105.9	109.3	112.4	114.0	121.3	–	–
Italy	100	104.2	108.9	108.9	101.7	95.1	99.2	–	–
Canada	100	106.2	103.7	104.0	103.6	102.7	102.6	102.3	–
USA	100	98.5	96.0	96.1	94.1	91.5	95.2	97.0	–

Source: Mouritzen and Nielsen (1988: 39).

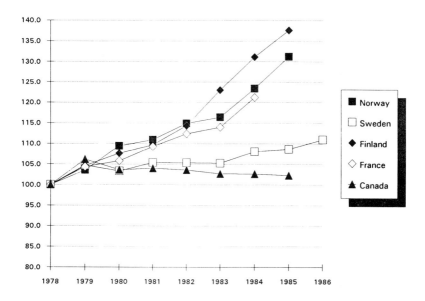

Figure 4.1   *Fiscal slack in five least stressed countries (data from Table 4.1)*

Comparing the situation in 1978 with that in the mid-1980s, the 10 countries can be categorized into three groups. Five countries – Finland, Norway, France and, to a lesser extent, Sweden and West Germany – are clearly better off at the end of the period than at the beginning. Four countries – Canada, Denmark, Italy and the USA – operate at about the same level at the end as at the start (100 ± 3). One country – the UK – is clearly worse off at the end of the period.

The patterns of change seem to follow the same lines. Finland, Norway, France and to some extent Sweden underwent what came close to a linear trend, a gradual expansion of the economic opportunities for municipalities. The remaining countries experienced major declines around the beginning of the 1980s but started to catch up around 1982–83. The only exception to this pattern is Denmark, where the deterioration of financial conditions started around 1982–83.

Whether or not one finds it appropriate to talk about a local fiscal crisis in a country depends on the period considered. Except for Denmark, the Nordic countries were definitely not in a position, at any time during the period, which justifies the term 'fiscal crisis'. This is also true for France.

Excluding the years before 1982, Danish municipalities were under pressure. For Denmark this crisis was of the same magnitude as that

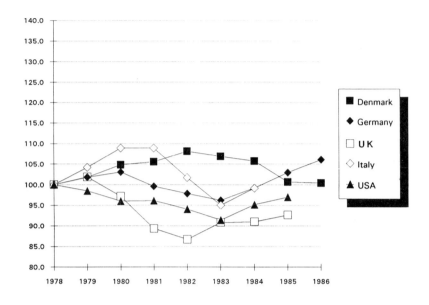

Figure 4.2   *Fiscal slack in five most stressed countries (data from Table 4.1)*

experienced by American municipalities in the period 1978–83 and West German local governments in 1980–83. The largest economic downturns are found in UK local government in 1979–82 and in Italy, where municipalities were brought into a shock condition during the two years 1981–83.

In the UK all indicators used by the Volkswagen group showed a worsening situation in the 1970s. By any measure that could be conceived of 'short of bankruptcy itself, communal finances in Italy [were] in a dire state' (Newton, 1980: 16). For the UK and Italy these negative trends evidently extended into the 1980s. From 1981 to 1983, for example, Italian municipalities lost almost 13 percent of revenues necessary to maintain existing service levels with constant rates of taxation.

Danish municipalities which were found to be in an excellent financial condition during the 1970s experienced some hardship beginning in 1983. However, up to that point the statement made by Newton and associates seems to be valid: 'Danish municipalities are generally in the enviable state of having a more than adequate income to cover their current and capital expenditure' (Newton, 1980: 13).

The Volkswagen group found no serious problems in Norwegian municipalities, although 'they seem to be getting worse rather than better' (Newton, 1980: 15). This prediction can now be definitely refuted:

Norwegian local governments enjoyed excellent financial conditions during the 1980s.

The relative ranking of municipalities in Sweden and West Germany seems to be the same in the 1980s as in the previous period.

After this brief journey back to the 1970s, we return to a more detailed description of some of the main trends in local finances in the 10 countries.

EXPLAINING SLACK AND STRESS

The score on the *SLACK* index is a function of four components: the tax base, grants, needs, and inflation. These components are in focus when we search for the immediate causes of fiscal stress (see Figure 4.3). Changes in the local tax base, grants, needs, and the rate of inflation arise from what we here term the 'underlying causes'. We first focus on the immediate causes.

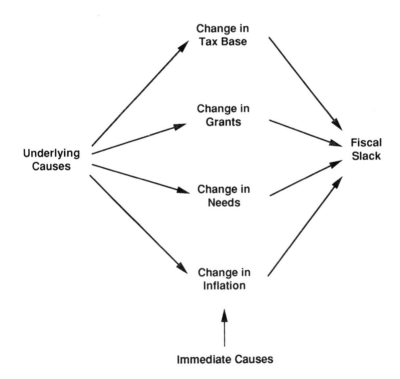

Figure 4.3   *Immediate and underlying causes of slack in local government*

Table 4.2   *Immediate causes of fiscal slack*

	SLACK index	SLACK	Contribution from		
			Tax base	Grants	Needs
Denmark					
1979–85	99.3	–0.7	4.7	–4.8	–0.6
1979–82	106.0	6.0	1.4	5.2	–0.5
1982–85	93.7	–6.3	3.3	–9.4	–0.1
Norway					
1979–85	126.9	26.9	14.8	14.0	–1.9
1979–82	111.0	11.0	6.1	5.9	–1.1
1982–85	114.3	14.3	7.9	7.3	–0.9
Sweden					
1979–85	103.7	3.7	2.0	2.4	–0.7
1979–82	100.6	0.6	–0.2	1.2	–0.4
1982–85	103.1	3.1	2.2	1.1	–0.2
Finland					
1979–85	131.6	31.6	20.9	13.6	–2.8
1979–82	109.3	9.3	6.6	4.0	–1.2
1982–85	120.4	20.4	13.3	8.8	–1.6
Germany					
1979–85	101.1	1.1	2.5	–2.0	0.7
1979–82	96.1	–3.9	–2.1	–1.5	–0.3
1982–85	105.4	5.4	5.0	–0.6	1.0
UK					
1979–85	91.0	–9.0	1.8	–10.0	–0.8
1979–82	85.1	–14.9	–2.2	–12.5	–0.1
1982–85	107.3	7.3	5.3	2.6	–0.7
France					
1979–84	116.1	16.1	6.3	12.4	–2.4
1979–82	107.6	7.6	1.3	7.9	–1.6
1982–84	108.0	8.0	5.0	3.9	–0.9
Italy					
1979–84	95.2	–4.8	0.4	–4.0	–1.2
1979–82	97.6	–2.4	0.1	–1.9	–0.6
1982–84	97.6	–2.4	0.4	–2.2	–0.6
Canada					
1979–85	96.0	–4.0	5.4	–2.8	–6.6
1979–82	97.4	–2.6	2.5	–1.2	–3.8
1982–85	98.4	–1.6	3.1	–1.7	–2.9
USA					
1979–85	98.3	–1.7	8.1	–3.2	–6.6
1979–82	95.4	–4.6	1.0	–4.8	–0.7
1982–85	103.0	3.0	7.4	1.4	–5.8

Source: Calculations based on data from Mouritzen and Nielsen (1988).

The immediate causes

In Table 4.2 the contribution coming from the three elements has been calculated for the 10 countries for three periods: 1979–85, 1979–82 and 1982–85. The idea is that score on the *SLACK* index (or rather, the difference between actual score and 100) can be decomposed into three parts showing the contribution from the tax base, grant and needs elements. (We omit the inflation component, as all calculations are made in fixed prices.) Take the Danish score of 106 for the period 1979–82. This score shows that local governments with a constant rate of taxation would have had a growth in revenues of 6 percent more over the period than what was needed to maintain existing service levels. The 6 percent is then decomposed into three parts. The main reason why municipalities in Denmark were better off in 1982 than in 1979 was the increase in grants to local governments, which contributed to the improvement in overall conditions by 5.2 percent. The tax base made a minor contribution of 1.4 percent, while increases in needs made a negative contribution of 0.5 percent.

About half of the 30 observations in Table 4.2 are indicative of growing resource scarcity. Only in a few instances do trends in the local tax base contribute to fiscal stress (3 out of 30 observations). In contrast, the needs element contributes to stress in almost all cases, albeit only modestly. It is evident that the major cause of fiscal problems in local governments is to be found in government grant policies. In about half of the cases, grants contribute to fiscal stress. In other words, local government fiscal problems seem to have their major root in reductions in grants to local governments.

A closer analysis of the data in Table 4.2 shows that grant policies in general reinforce rather than alleviate the effects that changes in the private sector economy (local tax base) have on the overall financial situation of localities. A low score on the tax base element thus lives side by side with a low score on the grant element.[2] This seems to be in stark contrast to the situation in the earlier period, where grants to local governments were widely used by higher levels of government in order to deal with urban fiscal problems (Wolman and Goldsmith, 1987: 171).

We now turn to a discussion of the major trends in the individual countries, with special emphasis on government policies directed towards local government finances. Unfortunately, information has not been available for West Germany and Canada.

Denmark

To the extent that there were fiscal problems in Danish municipalities before 1982, they had their root in a national recession, that is, in the local tax base: from 1978 to 1981 personal income decreased in real terms by almost 5 percent (Mouritzen and Nielsen, 1988: 46). However, the tax base of local governments rose slightly in the same period (Mouritzen and Nielsen, 1988: 29). This is due to the income policies pursued by the central government, which led to lower wage increases for government (including local government) employees than for private sector employees. In other words, the municipal cost index rose at a slower rate than the rate of inflation.[3]

In this period central government, led by the Social Democrats, implemented a relatively 'soft' policy of recommended or non-mandatory ceilings on the growth in local expenditures. The aim of government policies in this period was mainly to keep down local taxes.

In the fall of 1982, a conservative coalition government took over from the Social Democrats. The new government drastically changed policies towards local governments. In order to reduce the deficit in the national budget (in 1982 close to 10 percent of GDP: Mouritzen and Nielsen, 1988: 49), it implemented a series of cuts in general grants to local government. When the national economy entered a boom period in 1983, the conservative government introduced what was to become known as the 'take-home' principle. The argument was that the boom in the private sector was due to the economic policies of the government; consequently, the government had the right to 'take home' the resources that would otherwise flow into the local government sector. The policy instrument for doing this was immediately at hand: grant reductions. This policy was clearly based on experiences in the 1970s, which had shown central policy-makers that 'free-floating' money would never lead to lower local government taxes but rather to higher expenditures. (Cf. also the finding of the Volkswagen group cited earlier.)

The ways in which these grant reductions were distributed among local governments have differed since 1982. From the start, a principle was used according to which the rich localities (in terms of local tax base) were hit by the most severe cutbacks. After a couple of years, reductions were distributed according to *change* in the local tax base. Since 1986, however, the most visible and politically debated method has been to reduce grants to local governments that do not meet the still existing spending ceilings which for some years have aimed at zero growth. This method corresponds very much to the principles used by the UK Conservative government.

Norway

Drastic changes in Norwegian society have taken place during the last decade as a result of the 'oil boom'. In real terms, the gross domestic product rose by almost 32 percent from 1978 to 1986 (Mouritzen and Nielsen, 1988: 47). This has allowed Norwegian central government to continue a policy introduced at the beginning of the 1970s which heavily supports local governments through expanded categorical grants. In real terms, grants have increased by almost 50 percent from 1978 to 1985 (Mouritzen and Nielsen, 1988: 28).

The expansion of local government programs during the 1970s was, to a large extent, financed via loans. Because of rising interest rates, Norwegian local governments were hit by a mild 'crisis' around 1980. In several municipalities planned investments were dropped, and many municipalities successfully turned to the central government for help and were able to maintain their expansionist style by defining their objectives in a manner that corresponded to central government preferences. (In other words, they were able to get more grants.) It is, however, symptomatic for the situation in Norwegian local government that the crisis manifested itself as a reduction in the growth rate from 7–9 percent per year in the mid-1970s to around 2–3 percent per year at the beginning of the 1980s.

Sweden

Quite contrary to Norway, the Swedish economy was severely hit by the increase in oil prices and the general recession in the mid-1970s. The expansion of local government continued, however, not least through a further development of child-care programs and old-age policies. The local tax rate increased in most municipalities, especially in 1978. The parliamentary decisions on the above-mentioned welfare programs also led to an increase in the conditional grants in the early 1980s. In the same period a decision was made to increase the 'tax-equalization' grant, which is a general grant. Hence most of the municipalities went into the early 1980s with a fairly good economy. The national picture, however, was quite another story. The budget deficit was rapidly growing. The national government wanted, therefore, to reduce public spending in general. Some minor cuts were made in the grants, but on the whole no heavy cuts were made.

The general political pressure from the national government to reduce public spending made most local governments cautious. The risk of being hit by cuts in grants later reduced their willingness to increase local expenditures. A combination of a reduced increase in spending, surpluses arising from tax increases in the 1970s and increases in grants from central government produced a very favorable situation for many local governments in the beginning of the 1980s.

In the mid-1980s some reductions in grants were made, as well as some general changes in the grant system; but on the whole there was no severe fiscal stress among the Swedish municipalities.

Finland

Finnish local governments experienced an extremely favorable situation during the period under investigation. This was the product of a booming national economy. From 1978 to 1986 the gross domestic product increased by almost 30 percent in real terms (Mouritzen and Nielsen, 1988: 47). During the same period central government operated a budget that was, for all practical purposes, balanced. The debt of the government is the lowest found in any of the 10 countries (Mouritzen and Nielsen, 1988: 51).

These favorable external conditions left their traces on the fiscal well-being of local governments. The local tax base, as well as grants to local government, were exploding at a rate found in no other country. Consequently, municipal service production expanded dramatically, especially in the sectors of education, health care and social welfare.

From the mid-1970s, the Finnish central government and municipalities made agreements (covering one or two years) concerning the general level of increase in local expenditures, personnel and taxes. During the following years, the recommendations were followed by the local decision-makers, but later, expenditure increases have been higher than those agreed upon. From 1986 on, the local finance situation has worsened.

United Kingdom

For more than a decade, the status of local governments in Britain has steadily worsened, with consistent central government pressure being exerted on local authorities to reduce their expenditures and to limit increases in local tax rates.

What has been involved is a steady decline in the financial discretion of local authorities to determine for themselves both the levels of expenditure and the levels of revenue, especially for those that have exceeded the central government targets set for expenditure. This process of expenditure reduction and revenue control had been going on since the mid-1970s, but it intensified under the various Conservative governments of the 1980s. In fact, the biggest contribution to fiscal stress from decreases in intergovernmental grants is found in the UK, particularly in the first few years of the Thatcher government (cf. Table 4.2).

The main changes have followed from changes in the grant system and the tax system introduced in the early 1980s. Central government used its power to limit expenditures, to impose sanctions in the form of grant reductions for overspending, and to impose limits on the tax-raising capacity of those authorities that persistently overspent.

France

Several factors explain the level of fiscal slack in France. The proportion of local spending remains lower than in most other countries; therefore control of local expenditures is less crucial here than in other countries for the success of a policy of austerity and taxation reduction. But in fact, central government policies did not lead to any deliberate attempt to reduce taxes and public spending, as was the case in several Western countries.

Municipalities in France have to bear only a small share of welfare expenditures linked to economic recession and unemployment. A national institution manages the obligatory social security contributions – paid both by employers and employees – and allocates them (pensions, health insurance and various other allowances). Most additional social welfare expenditures are managed by other local governments, the *départements*, which receive an annual subsidy from the municipalities.

From 1983 onwards the communes' fiscal situation changed. The socialist central government began implementing its policy of austerity and announced its intention of reducing income taxes, as they were considered an impediment to restructuring the economy. This objective was pursued even more radically by Jacques Chirac's government from 1986 on, but that did not entail any measures of constraint as far as the communes were concerned. The central government, faced with the powerful Association of French Mayors, could do no more than make verbal statements recommending moderation in local expenditures if the policy of decentralization was to succeed.

Italy

In 1973 the Italian parliament approved a tax reform that made local finances highly dependent on central grants, which now account for more than 50 percent of total local revenues (Mouritzen and Nielsen, 1988: 33). A peculiar system of revenue sharing was introduced which gave to each municipality an annual grant based on the lost revenues from the reduced taxes.

Owing to exploding local government debt, the Italian government in 1977 intervened and took over the burden of debt service. But this favor had its price. The government also imposed ceilings on the annual growth in expenditures. A ceiling was fixed for each municipality on the basis of the previous year's levels (with redistribution criteria favoring southern and poor cities). A grant was then allocated to each locality equal to the difference between allowed expenditures and the income from all other sources. This system of financing discouraged the effort of local governments to improve services through increases in taxes and user charges, because any increase would be eaten up by a corresponding grant reduction.

In the beginning of the 1980s a completely different system was implemented, which in fact turned things around. Grants were distributed according to local tax base and fiscal effort; and local governments were encouraged to increase tax rates and user charges, and to introduce new taxes.

Looking at Table 4.2, it is clear that the major cause of fiscal crisis in Italian local governments is the grant component. The annual increase in grants has proven to be significantly lower than the rate of inflation.

USA
From the late 1970s to the mid-1980s, several major events occurred in the USA that affected local finances. These changes involved changes in the private economy, as well as in federal and state fiscal policies.

Certainly, the major event was the national recession in 1981–82, which was quite different from other post-war recessions in its severity and its effect on many medium-sized and large cities. In 1981–82, the rate of unemployment in the USA rose from 7.6 to 9.7 percent (Mouritzen and Nielsen, 1988: 48), which reflects many business closings, mostly in north-eastern and mid-western cities.

The Reagan administration moved to cut back federal aid programs to cities, thereby triggering some of the austerity in later years. Although local governments receive a significant proportion of their intergovernmental revenues from state governments, many states were hit by the recession also and moved to reduce their support of the cities.

In the period 1979–82, local governments in the USA were hit by fiscal stress almost solely because of the grant policies of the federal government and state governments (cf. Table 4.2). As a result of the recession, the local tax base was not able to compensate for these losses. From 1982 onwards the local tax base increased, as did grants to local governments. Both these elements contributed to a more favorable position for local governments.

An important aspect of the US picture is tax restrictions. Following the Proposition 13 movement in California in 1978, there has been a more visible attitude toward restrictions on local taxes, most notably property taxes. Several states, such as Michigan and Massachusetts, introduced some form of restrictions on expenditures, taxes or other fiscal variables. In general, local political leaders encounter much closer supervision by the electorate on revenue enhancements, although there are often ways to avoid the restrictions.

Some underlying causes
In seeking explanations of fiscal stress in local government, most studies have focused on individual cities within one country. These studies emphasize explanations such as demographic changes, the exodus of people from central cities, structural changes in the national economies which may result in some regions becoming less competitive (Rubin,

1982), a declining economic base, inflation and rising public employment costs (Bahl et al., 1978).

For the public choice approach, the root of urban fiscal problems is the asymmetric decision-making process in which interest groups and bureaucracies make excessive demands for more and better services. Because of poor information and the relative invisibility of taxes, specialized groups are able to dominate the decision-making process. Benefits are concentrated while costs are shared among all taxpayers (Kristensen, 1980; Rubin, 1985). This approach focuses primarily on the responses of local governments to external conditions, as it predicts that expenditures will tend to grow faster than the private economy can support. In other words, the explanatory power of the public choice approach is dependent on how fiscal problems are conceptualized. If one applies a maladaption concept of fiscal crisis (cf. Chapter 3 above), public choice arguments may be highly relevant. Such arguments make us aware of possible between-country institutional differences, which – acting through their effect on the incentive structures of citizens, groups and political leaders – may have important effects on the way cities respond to fiscal stress in different countries. (The responses are analyzed in Chapters 8, 9 and 10.)

Here we have chosen a socio-economic imbalance concept of fiscal stress, where structural theories have more to offer in a comparative context. This perspective stresses the combination of business dominance over the fiscal policy-making process and the flight of middle-class taxpayers from the central cities as a cause of fiscal stress (Rubin, 1985). To this extent, its explanatory power is to be found in within-country analyses. However, a central theme of the structuralist perspective is the dependence of local governments on the 'taxable revenues generated by private investment' (Piven and Friedland, 1984: 406), that is, the link between developments in the private sector and the fiscal well-being of governments. According to Rubin (1985: 475), it was predicted by the neo-Marxists (particularly O'Connor) that the central governments would not sustain the level of support to local governments via grants. The crisis of the welfare state would be *exported* from one level of government to the other.

In the remainder of this chapter we will take a first look at the possible links between national economic growth, the crisis of the welfare state (central government) and the fiscal well-being of local governments. Omitting the needs component in our fiscal stress definition, it is clear that the effect of the other two components – the local tax base and grants – is of a different character. Downturns in the national economy are more or less automatically transformed to fiscal stress through changes in the tax base, although with different intensities and time lags depending on the specific character of the taxation system. However, to the extent that grants are affected by the conditions of the national economy, this effect is mediated through deliberate government actions: central governments may reinforce or alleviate the effects of the private economy through its grant policies. In this chapter, however, we merely look at total effects; in the next chapter the indirect effects via the grant policies of central government are analyzed in more detail.

The analysis is based on pooled data, where the unit of analysis is yearly observations for 10 countries over the period 1978–86. The dependent variable is change in fiscal slack from year $t-1$ to year t calculated on the basis of the figures in Table 4.1.[4] Slack is expected to depend on three sets of factors: (1) ideology of central government; (2) health of the national economy; and (3) the fiscal health of central government. Because of the particular data matrix, it is not possible to estimate the effect of these sets of factors in a multivariate context, and as a consequence we merely look at the bivariate correlations.[5]

Ideology of central government is simply measured as party color of government using the traditional left–right distinction in each country. Three indicators measure the health of the national economy: real growth in GDP from year $t-1$ to year t, change in balance of payment surplus from $t-2$ to $t-1$, and balance of payment surplus in year $t-2$. The fiscal condition of central government is measured through four indicators: change in budget surplus and government debt from $t-2$ to $t-1$ and the actual budget surplus and government debt at $t-2$. Balance of payment surplus is measured as a percentage of GDP, while central government debt and budget surplus are measured as a percentage of public expenditure.

The idea behind the indicators of health of the national economy and central government is that they are known criteria on the basis of which central government will decide on its policies toward local government, thereby affecting the degree of slack. One notices immediately that some of the indicators measure changes while others measure levels. The change variables are lagged one year while the level variables are lagged two years.[6] The analysis reveals that slack in local government seems strongly related to some factors but not to others; cf. Table 4.3.

Apart from growth in GDP, the change measures are not strongly related to the degree of slack in municipalities. In contrast, the level indicators, which show the fiscal health of central government, are highly correlated with slack. It seems as if central government reacts more on levels than on changes in levels when it comes to exporting fiscal troubles downwards in the system. What would seem to be the most important of the health factors, after GDP growth, is the size of the government debt and the size of the national government surplus (or deficit). A small government debt and a large government budget surplus in year $t-1$ thus seem to lead to better conditions for local government in year t.

The bivariate analysis does not support the conventional wisdom that

Table 4.3 *Correlation between fiscal slack in municipalities and ideology of central government and fiscal health indicators* (*Pearson's* r)

	Correlation	Significance
Ideology		
Left-wing govt.	0.02	0.45
Health of national economy		
GDP growth	0.52	0.00
Growth in balance of payments surplus	0.07	0.28
Balance of payments surplus	0.12	0.17
Fiscal health in government		
Growth of govt debt	0.18	0.09
Growth of govt surplus	−0.14	0.16
Govt debt	−0.42	0.00
Govt budget surplus	0.41	0.00

Number of cases from 54 to 71.

Source: Based on pooled data from Mouritzen and Nielsen (1988) covering the period 1978 to 1986.

localities have better conditions if central government is led by the leftist parties. However, we shall return to this theme in more detail in the following chapter, where the grant policies of governments are subjected to a more profound analysis.

Finally, we return to the decisive factor, change in the GDP. Our bivariate analysis shows that, among the 10 selected countries during the 1978–86 period, local fiscal conditions were fairly closely related to change in GDP indicated by a correlation coefficient of 0.52. The relation was however very different from country to country. (Cf. Table 4.4, where we have ranked the countries according to the size of the correlation coefficient.)

The fiscal health of municipalities is most strongly related to growth in GDP in Italy and the UK, less so in West Germany and the USA. Norway, Sweden and France and to a lesser extent Canada exhibit only moderate correlations, while in Denmark and Finland we find negative correlations. The negative relation in Denmark has to do with the fact that the Social Democratic government, which was in power to the end of 1982 (a period of economic decline), was very reluctant to reduce grants to local government. This was one of the reasons why the conservative coalition government which was in power in the following boom period fought a battle against a large budget deficit, the major weapon of which was reductions in grants to local government. We return to these reductions and the effects of party politics in Chapter 5.

Table 4.4 *Correlation between fiscal slack in municipalities and change in gross domestic product, 1978–86*

	Correlation	N
Italy	0.94	6
UK	0.93	7
Germany	0.79	8
USA	0.72	7
Norway	0.52	7
Sweden	0.43	8
France	0.30	6
Canada	0.23	7
Finland	–0.07	7
Denmark	–0.53	8

Source: Calculations based on data from Mouritzen and Nielsen (1988).

Apart from the negative figure for Denmark (and the close to zero figure for Finland), the general impression is that fiscal slack or its opposite, fiscal stress, to a large extent has its roots in developments in the national economy: the fiscal crisis of the state is exported to local government.

ARE THERE DIFFERENCES BETWEEN CITIES?

Aggregate analyses are always crude because they do not reveal differences within a country. There are great variations in the fiscal conditions of municipalities, and these variations also seem to be a product of the country on which we are focusing.

Table 4.5 shows descriptive statistics for five countries based on city-level data. The fiscal *SLACK* index is, in principle, calculated as the aggregate index studied earlier. First, we observe no relation between mean score and variation since we find the lowest coefficient of variation for Sweden, which has an average score on the *SLACK* index. Further, the largest dispersion on the *SLACK* indicator is found in the USA, which experienced a relatively high degree of stress. But Denmark also experienced a high degree of stress. Here we find a fairly low degree of variation between municipalities, a reflection of the way grants to local governments are distributed as a function of the local tax base.

Table 4.6 shows, for cities in the same five countries, the correlations between fiscal slack and wealth, size and socialist power. Based on the five-country sample, we find no pattern with respect to slack and equality. It has often be suggested that fiscal stress in local government would lead to larger inequalities between cities, but this is not true. For three

Table 4.5 *Variations in fiscal slack in municipalities*

	Denmark	Norway	Sweden	Finland	USA
Mean	92.8	116.9	106.4	121.9	95.3
Coeff. of variation	0.06	0.12	0.03	0.06	0.20
Frequency distribution					
Above 120	0.0	26.7	0.0	44.0	5.6
110–120	0.4	49.3	9.0	56.0	8.7
100–110	5.4	22.4	89.2	0.0	20.0
90–100	68.4	1.3	1.8	0.0	30.7
80–90	23.6	0.0	0.0	0.0	19.2
Below 80	2.2	0.2	0.0	0.0	15.8
Total	100.0	99.9	100.0	100.0	100.0
N	275	450	279	84	850
Period investigated	1982–86	1981–84	1982–85	1981–85	1980–84

countries, the richer cities seemed to be better off. However, this is true for the two slack countries, Norway and Finland, as well as for the USA, where local governments were in a stress situation. In Denmark the trend goes in the opposite direction. This is due mainly to a very strong movement of industry and jobs from the rich, metropolitan and

Table 4.6 *Correlation between city wealth, city size and socialist power and fiscal slack in municipalities (Pearson's r)*

	City wealth[1]	City size (log)	Socialist power[2]
Denmark	−0.57	−0.27	−0.29
	(0.00)	(0.00)	(0.00)
Norway	0.22	−0.18	−0.12
	(0.01)	(0.01)	(0.01)
Sweden	−0.14	−0.08	0.02
	(0.01)	(0.08)	(0.38)
Finland	0.24	0.10	−0.07
	(0.00)	(0.19)	(0.25)
USA	0.20	−0.06	0.01
	(0.00)	(0.03)	(0.35)

[1] City wealth is measured as taxable (or personal) income per capita in the first year of the period investigated (cf. Table 4.5, last row).
[2] Socialist power is percentage of seats in city council held by traditional left-wing parties (in the USA, the Democratic party).

Source: Data collected by Fiscal Austerity and Urban Innovation project national teams.

eastern part of the country to the traditionally poor western periphery. In Sweden the poorer localities also do better.

It is probably the general impression in most countries that the big cities are hit relatively hard. This impression gains some credibility, as slack tends to be lower the bigger the local government is in terms of population. Only in Finland do we find the opposite, albeit weak, tendency.

In two countries, Denmark and Norway, we find a tendency for socialist municipalities to be hit relatively harder than municipalities where the right-wing parties are strong. This tendency, which may be a mere reflection of the size of the city, will be analyzed in greater detail in the next chapter.

SUMMARY

Local officials in most countries always seem to be complaining about their fiscal conditions. It is always easy to explain such complaints. They are to a large extent caused by the nature of the political process. However, they are not always easy to justify. In this chapter we have shown that, in the aggregate, local officials in four countries, Norway, Finland, Sweden and France, have lived under extremely favorable conditions during the period investigated. In the other six countries, they have for shorter or longer periods been under fiscal pressure. This is particularly true for Italy and the UK and to a lesser extent for Denmark, West Germany and the USA. Canadian municipalities have lived under rather stable conditions over the period 1978–86 although with a weak downward trend.

It was also shown that the fiscal well-being of municipalities is closely linked to the policies pursued by central governments. In most cases of fiscal stress, the major cause is to be found in reductions in grants going to municipalities. In contrast to previous periods, central governments' grant policies were not aimed at alleviating the effects of the private sector economy: rather, grant policies seemed to reinforce these effects.

There was a strong tendency for the fiscal constraint on municipalities to reflect the general trends in the national economy and the fiscal health of the central government as measured by government debt and budget surplus. Thus central governments, to a large degree, seem to export their financial problems to local governments. The mechanisms of grant policy are the subject of the next chapter.

NOTES

1 Instead, we could have taken the rate of taxation from the preceding year. However, this measure is very sensitive to changes in level of taxation the year before. If for instance local governments in a country had responded

to fiscal stress in 1979 by massive increases in tax effort, the score for 1980 would merely be a reflection of this policy response and not of changes in the external conditions.

2 In a regression with two observations per country (1979–82 and 1982–85) R^2 is 0.36 and the slope is 0.88 (with change in grants as the dependent variable). In other words, each time the tax base element contributes to fiscal stress by a value of –1, the contribution from the grant element is –0.88.

3 The consumer price index rose from 100 to 138 while the municipal price deflator rose from 100 to 132 (both from 1978 to 1981); cf. Mouritzen and Nielsen (1988: 43 and 48).

4 Cf. note 1, which justifies this procedure.

5 The problems are discussed in more detail in Chapter 5; cf. particularly n. 1.

6 Other laggings were tried. The ones chosen produced the largest correlation coefficients.

REFERENCES

Bahl, Roy, Bernard Jump and Larry Schroeder 1978. 'The Outlook for City Fiscal Performance in Declining Regions', in Roy Bahl (ed.), *The Fiscal Outlook for Cities: Implications of a National Urban Policy*, Syracuse NY: Syracuse University Press.

Kristensen, Ole P. 1980. 'The Logic of Political Bureaucratic Decision Making as a Cause of Governmental Growth', *European Journal of Political Research*, 8: 249–264.

Mouritzen, Poul Erik and Kurt Houlberg Nielsen 1988. *Handbook of Comparative Urban Fiscal Data*, Odense: Danish Data Archives, University of Odense.

Newton, Kenneth 1980. *Balancing the Books: Financial Problems of Local Government in West Europa*, London: Sage.

Piven, Frances Fox and Roger Friedland 1984. 'Public Choice and Private Power: A Theory of Fiscal Crises', in Andrew Kirby, Paul Knox and Steven Pinch (eds), *Public Service Provision and Urban Development*, New York: St Martins Press.

Rubin, Irene 1982. *Running in the Red: The Political Dynamics of Urban Fiscal Stress*, New York: State University of New York Press.

Rubin, Irene 1985. 'Structural Theories of Urban Fiscal Stress', *Urban Affairs Quarterly*, 20: 469–486.

Wolman, Harold and Michael Goldsmith 1987. 'Local Government Fiscal Behavior and Intergovernmental Finance in a Period of Slow National Growth: A Comparative Analysis', *Environment and Planning*, 5: 171–182.

5

Fiscal Stress and Central–Local Relations: The Critical Role of Government Grants

Harold Wolman, with Michael Goldsmith,
Enrico Ercole and Pernille Kousgaard

When the Thatcher government took power in 1979, central government was providing 53 percent of the net expenditure of the major metropolitan local authorities in England and Wales. Faced with a severe economic downturn and pressure on central government budgets, only seven years later the government contribution had dropped to 46 percent. Many of the large urban authorities – almost all of which were run by Labour-controlled councils – experienced severe fiscal problems as a result of the decline in grant as well as other central government measures designed to reduce local government revenue raising and spending. Liverpool's dilemma, greatly exacerbated by the political animosity between the far left Labour-controlled council and the Tory government, is the best known and documented case of the fiscal chaos that resulted.

The British experience seems to confirm popularly held beliefs as well as much conventional wisdom among social scientists: a central government will seek to export its fiscal problems to lower-level governments, and this is particularly true for conservative governments critical of an ever growing public sector. In this process of retrenchment, the government will seek to reward its allies and punish its opponents.

In this chapter we question whether this conventional wisdom is generally true for all countries. In some respects the British experience can be considered a deviant case; in other respects we are able to find similar patterns across countries; finally, in certain respects we are able to qualify the observations made from the British case.

In the course of addressing these questions we test five propositions:

1 The fiscal crisis of the state is exported to local governments; that is, when the national economy is in bad shape local governments will come to bear more than their share of the burden.
2 The extent to which this happens is, however, conditioned by

(a) ideology: local governments will suffer most under conservative regimes, less under leftist governments;
(b) nature of the grant system: a highly consolidated grant system will reduce the political costs for a government reducing support to localities. A highly fragmented grant system, on the other hand, will produce large political costs if grants are reduced.
(c) dependency: the more local governments rely on grants as a source of revenue, the more likely it is that they will come to suffer from fiscal stress in times of crisis in the national economy.
3 The distribution of rewards and punishments to local governments follows partisan political lines: a conservative government will particularly reduce financial support to leftist local governments, whereas a leftist government will make sure that conservative local governments are hit hardest.

We proceed by first examining the extent and causes of grant change during the 1978–86 period. As Chapter 4 established, change in the amount of grant to local government is the most important 'immediate' factor in bringing about local fiscal stress. Fiscal stress, as measured by the fiscal *SLACK* score (see Table 4.2 above), consists of a composite of changes in three factors: tax base, grant and population (a proxy for change in needs). The simple correlation between grant change during the previous year and fiscal stress (measured from the previous year) for the local government systems in each of the 10 countries over 1978–86 was $r = 0.89$, while the correlation between tax base change and fiscal stress was 0.54 and between changes in population and fiscal stress, only 0.02. It is clear that grant change played the dominant role. In this first section we study the relationship between grant policies, national economic growth, color of national government and characteristics of the grant system. Here we use national aggregated data and pooled data.

Next, we consider the distributional impact of grant change among local governments. We will focus, in particular, on the distribution of grant change among local governments and will examine the extent to which grant distribution is based on party-political considerations as opposed to objective criteria. This section is based on city-level data from seven countries.

The chapter then turns to a consideration of the extent to which city governments replaced lost grants with own-source resources. Of particular interest here is the extent to which replacement behavior is contingent upon the health of the national and local economy. We analyze these questions using national data as well as city-level data. The latter, unfortunately, cover only two countries.

The chapter concludes by examining the impact of grant change on several important aspects of central–local relations including city government vulnerability, autonomy, equalization, and the concentration/dispersal effects of grant policies on municipal governments. Here too, we apply national as well as city-level data.

In the course of this examination, we attempt to determine whether the various countries and their systems of local government exhibited similar patterns, and, to the extent they did not, we attempt to account for differences among the countries. Differences might be due to a variety of factors, including variations among the countries in:

- national economic growth (that is, did slow growing countries differ from fast-growing ones?)
- ideology of national government (left, right)
- direction (and extent) of grant change (increase, decrease)
- type of grant system (consolidated, fragmented).

LOCAL GOVERNMENTS IN A TURBULENT NATIONAL ECONOMY

The extent of real grant change
How substantial was grant change over the period studied? Over the course of the full period (1979–85), the average change in real grant was an increase of 7.2 percent. However, as Table 5.1 indicates, this average obscures a substantial diversity in experience. In Norway and Finland real grant increased by more than 40 percent during this period, while in the UK and Denmark it fell by more than 10 percent. (Indeed,

Table 5.1   *Real grant change*

	1979–82 (%)	1982–85 (%)	1979–85 (%)
Denmark	12.0	–21.0	–11.5
Norway	18.8	21.4	44.4
Sweden	3.5	3.0	6.6
Finland	12.9	28.3	44.9
Germany	–3.1	–1.2	–4.2
UK	–18.9	4.6	–15.1
France	15.5	7.5	24.3
Italy[1]	–2.1	–2.5	–4.5
Canada	–2.1	–3.2	–5.2
USA	–10.3	3.3	–7.2
Mean	2.6	4.1	7.2

Unit of analysis: countries.
[1] Italian data were not available for 1985: instead, 1984 data are used.

Source: all tables in this chapter derive from data in Mouritzen and Nielsen (1988).

real grant actually declined in 6 of the 10 countries over the 1979–85 period, increasing only in Sweden and France in addition to Norway and Finland.)

Grants also increased much more rapidly in the expansionary (1982–85) period than in the earlier slow-growth (1979–82) period. The average change in real grant was 4.1 percent in the expansionary period compared with 2.6 percent in the slow-growing years. Indeed, in some countries grant change varied dramatically between the two periods. In the UK, for example, grants fell by 18.9 percent in real terms between 1979 and 1982, then increased by 4.6 percent between 1982 and 1985; in the USA they fell by 10.3 percent during the first period, then increased by 3.3 percent in the second. In Denmark, real grant rose by 12.0 percent between 1979 and 1982, then fell by 21.0 percent between 1982 and 1985.

Causes of grant change
What is it that brings about changes in the amount of grant flowing to local governments? To what extent do grant changes simply reflect the economic environment a country faces, the fiscal health of national governments, and the political preferences of the party (or parties) controlling national governments?

We first turn to the relationship between the national economic environment, as measured by the rate of growth in gross domestic product, and grant change. Over the entire period of 1979–85, percentage changes in GDP explained 64 percent of the variation in percentage changes in grant for the 10 countries, based on a simple bivariate relationship ($r^2 = 0.64$). During the 1979–82 period, which for most of the 10 countries coincided with a severe economic downturn, the explanatory value of GDP change was slightly lower ($r^2 = 0.53$), but during 1982–84, a period of economic recovery for most countries, the relationship completely disappears ($r^2 = 0.02$).

It thus appears that grant was responsive to GDP change during the period of slow economic growth or decline but not during periods of economic growth. It seems likely that during a period of relative economic decline national governments turn quickly to grants as a means of reducing or holding down national expenditures. Grants are more 'controllable' in this sense than income support, national security or other forms of national government expenditure. However, during times of relative economic expansion, grants are no more likely to be systematically increased than other forms of government spending. Thus, in some countries (Norway, Finland) grants expanded vigorously with GDP growth, while in others (Canada and Denmark) grants were stagnant or declined even while GDP was increasing.

The relationship between GDP and grant change is much stronger over a longer period (e.g., measuring the percentage changes in both variables

Table 5.2   *GDP and grant change*

	GDP and grant change, 1978–86 (r^2)	GDP growth, 1979–86 (%)
Denmark	0.403	20.3
Norway	0.021	34.9
Sweden	0.254	16.7
Finland	0	34.3
Germany	0.458	15.2
UK	0.557	12.6
France	0.070	15.4
Italy	0.864	21.4
Canada	0	25.0
USA	0.422	18.0

All figures and calculations based on real growth.
Unit of analysis: countries.

from 1979 to 1985) than on a year-to-year basis. Using pooled data (GDP change and grant change for each year between 1978 and 1986 for all 10 countries where data are available), the r^2 is only 0.13. This suggests that adjustments to grant systems caused by changes in the national economy occur over an extended period of time rather than instantaneously.

An analysis of the annual relationship between GDP change and grant change on a country-by-country basis provides additional insight. In five countries (Denmark, West Germany, the UK, Italy and the USA), GDP change accounted for a reasonable proportion of grant change ($r^2 = 0.40$ or greater), in one country (Sweden) for a modest amount ($r^2 = 0.25$) and in the four remaining countries (Norway, Finland, France and Canada), for essentially no change ($r^2 = 0.07$ or less). Closer inspection suggests that annual GDP change appears to have no relationship to annual grant changes in countries where GDP growth is high (the three countries with the highest rate of GDP growth – Finland, Norway and Canada – all exhibited no relationship between GDP change and grant change), but to have a substantial relationship where GDP growth is modest or low. (The r^2 between GDP change and grant change in the UK – the country with the lowest rate of GDP growth between 1978 and 1986 – is 0.56, and in West Germany – the country with second lowest rate of GDP growth – it is 0.46: see Table 5.2.) This re-enforces our earlier finding that grant is a very vulnerable form of governmental expenditure during periods of economic problems but is neither systematically favored nor discriminated against during periods of economic prosperity.

However, is grant equally vulnerable in all countries during periods of slow national growth? We might hypothesize that:

1 during periods of slow national economic growth, countries with more consolidated grant systems (cf. Chapter 1) will suffer greater grant reduction than countries with fragmented grant systems (proposition 2b);

2 during periods of slow economic growth, grant will be reduced more by national governments of the right than by those of the left (proposition 2a).

With respect to the first hypothesis, the logic is that in fragmented grant systems, which are characterized by a variety of categorical grants, interest groups concerned with specific program functions cluster around and protect each of the various programs, thus making them difficult to cut. General grant programs, on the other hand, have fewer defenders, since interest groups concerned with a specific program can have no assurance that the recipient will use the funds as they wish.

Data on general and conditional grants are available for only 7 of the 10 countries. If we take general grant exceeding 33 percent of total grant as indicating a consolidated grant system, then three countries (Denmark, the UK and France) had consolidated grant systems, while four (Norway, Sweden, Canada and the USA) had fragmented grant systems (see Table 5.3). During the 1978–86 period these 7 countries experienced a total of 19 years of relatively slow national economic growth or decline (defined as an annual growth rate in real GDP of less than 2 percent). Our hypothesis suggests that grant should have grown more slowly (or declined) during these slow-growth periods in countries with consolidated grant systems relative to those with fragmented systems. However, during these years the annual growth rate for real total grant averaged 1.40 percent for countries with consolidated grant systems and –0.16 percent for countries with fragmented systems, the opposite of the predicted pattern. Among the consolidated systems, Denmark and France had consistent increases in real grant during slow-growth periods, while

Table 5.3 *Degree of consolidation of grant system*

| | General grant as % of total grant | | |
	1979	1982	1985
Denmark	33.7	35.0	33.1
Norway	15.2	15.6	14.7
Sweden	9.8	16.6	16.0
UK	65.9	66.0	54.8
France	58.4	58.8	59.3
Canada	17.9	12.0	11.4
USA	26.7	24.4	25.4

Unit of analysis: countries.

the UK had declines during slow-growth periods, Sweden had very little grant change, and Canada and Norway both had one year of reasonable growth and one of decline.

If we take a stricter definition of consolidation of grant system by requiring that at least 50 percent of grants be provided in the form of general grant, then Denmark moves from a consolidated system to a fragmented system. The results do change somewhat; there is now virtually no difference in the average annual percentage change in real total grant during years of slow national economic growth between countries with consolidated grant systems (0.56 percent) and those with fragmented systems (0.73 percent). Expanding the criteria for slow economic growth to include all periods in which real GDP grew by 2.5 percent or less adds seven additional slow-growth years, but does not change the results.

The second hypothesis fares somewhat better. There were 27 years during the 1978–86 period during which the 10 countries had real GDP growth rates of 2 percent or less and 34 years in which real growth rates were less than 2.5 percent. Taking the latter figure as our standard of slow growth, real grant declined by an annual average of 2.24 percent in the 8 cases in which the national government was controlled by right-wing parties during the previous year (i.e., party control in year $t-1$ for grant change from year $t-1$ to year t) and by 0.66 percent in the 17 cases in which the national government was controlled by left-wing parties. It also increased by an average of 0.122 percent in the 5 cases in which national government control switched from right to left or left to right during the year and declined by an average of 0.66 percent in the 8 cases in which the government was controlled by a coalition.

Obviously, the number of cases is too limited and the differences between them are too small for us to reach any conclusive statement, other than that the hypothesis is not disproven and deserves further testing.[1]

THE DISTRIBUTION OF GRANT CHANGE: PARTY POLITICS OR 'OBJECTIVITY'?

What kinds of municipal governments were winners and what kinds losers in terms of changes in grant received? Changes in grant distribution can occur as a result of political considerations (national governments rewarding municipalities controlled by their parties) or as a result of the more inexorable workings (at least at first glance) of objective criteria in grant distribution formulas (for example, population, per capita income, tax base, needs factors such as number of children under the age of five, etc.). We say 'at first glance' because the objective criteria incorporated in grant formulas (that is, which factors; how weighted) may themselves be the result of political considerations present at the time the formula was written concerning who the formula would be likely to advantage and disadvantage. Moreover, grant formulas can be amended or new grants instituted to reward different sets of recipients. Thus, while it might seem that countries characterized by more fragmented grant systems, featuring large numbers of discretionary grants, would be more susceptible to the utilization of political criteria in terms of grant distribution, this may not necessarily be the case.

We test two hypotheses with respect to the distribution of grant change among local governments. The first hypothesis specifies that national governments reward their allies and penalize their opponents with respect to the distribution of changes in the amount of real grant (the 'partisan politics' hypothesis). The second hypothesis specifies that the distributional impacts of grant change among local governments is explained by the operations of objective need and resource criteria incorporated in grant formulas (the 'objective criteria' hypothesis). These criteria operate in predictable ways to advantage needy and low-resource municipalities during times of grant growth and to disadvantage them during times of grant cutback.

We will test these hypotheses by analyzing the distributional impact of grant change on municipalities classified by party control, city size, wealth, economic base change and economic distress. We will also examine whether the distributional patterns of grant change differ between countries with high and low increases in grant to local governments.

Party politics and grant policies
The partisan politics hypothesis assumes that national governments direct grant changes to the advantage of local governments whose voters are their partisan supporters and which are controlled locally by their party. Analysis of the distribution of grant change by party control yields very mixed results with respect to this hypothesis. Data are available for only seven countries. In Denmark the findings were quite contrary to expectations. Although the national government was controlled by the Social Democrats until 1982, the increase in grant over 1979–83 averaged 0.48 percent of the local tax base for municipalities controlled by the right compared with 0.21 percent for municipalities controlled by the left. From 1982 to 1986, under a Conservative government, grant to municipalities controlled by the right (presumably those closest politically to the national government) declined by an average of 4.51 percent of the local tax base, while grants to municipalities controlled by the left declined by only 4.00 percent of the local tax base.

Grant change by party in France conformed more readily to conventional expectations. In 1981 the Socialist Mitterrand government succeeded the conservative Giscard d'Estaing government. Between 1980 and 1984, grants to municipalities controlled by the left (in 1983) increased by 60.3 percent for Socialist-controlled municipalities and 66.0 percent for Communist controlled-municipalities, while grants to the right increased by 55.4 percent for Gaullist municipalities and 53.3 percent for UDF conservatives. In addition, there is evidence that the direction of party change in the 1983 municipal elections affected grant distribution in predicted ways. The 20 municipalities in which party control switched from left to right averaged grant increases of 64.8 percent between 1980 and 1984 compared with 69.3 percent for those municipalities that remained Socialist; the 12 municipalities with party control change from Communist to right averaged grant increases of 56.6 percent compared with 66.0 percent for those that remained Communist. This is consistent with Nevers's conclusion that the major French grant, the DGF, is 'rather highly related to the political colour of the municipalities' (Nevers, 1988: 34).

In Sweden, similar mechanisms appeared to be at work. Municipalities controlled by the right had real grant increases of 122.2 percent from 1978 to 1986 compared with 116.8 percent for those controlled by the left. However, this modest, but statistically significant, difference occurred wholly during the 1978–82 period when the Conservatives controlled the national government. Under the Conservatives a new grant reform took effect in 1980, guaranteeing all municipalities at least 103 percent of mean tax capacity (that is, supplementing local resources with national grants to ensure that all municipalities reached that level). As a consequence, national tax equalization grants were extended for the first time to more affluent, Conservative-controlled municipalities. During 1983–86, after the Social Democrats regained control of the national government, there was no significant difference in grant increases to municipalities by party color.

In Norway and the USA, the analysis found no significant differences in municipal grant change with respect to either party control or change in party control at the local level.

In Finland, which was controlled by a coalition government at the national level, there was essentially no difference in grant change over 1980–84 between municipalities that were socialist-controlled in both 1980 and 1984 and those that were socialist-controlled in 1980 but not in 1984. However, those municipalities controlled by non-socialist parties in both 1980 and 1984 averaged grant increases of only 36.9 percent compared with 42.0 percent for municipalities controlled by socialists in both years.

In Italy, which was also controlled by a coalition government, grants to the 101 municipalities controlled by the left in both 1980 and 1984 declined by an average of 3.1 percent in real terms over that period, while grants to the 113 municipalities controlled by non-left parties rose by an average of 3.7 percent. On the other hand, grants to the 15 municipalities that changed from non-left to left control between 1980 and 1984 increased by 13.2 percent, compared with a decline of 1.7 percent for municipalities where control switched in the other direction (left to non-left). The differences were significant.

Unfortunately, we did not have British city-level data available for this analysis. However, a recent study by Page et al. (1990) only weakly supported the partisan politics hypothesis. Even in the highly politicized period of 1981–88 there was only a weak relationship between grant change and the partisan composition of the council, although in the expected direction of favoring Conservative local governments.[2]

The above account suggests no obvious common pattern. France, Sweden and to a minor extent the UK behaved in ways that support the partisan politics hypothesis, but Norway, the USA and Denmark did not. (Indeed, Denmark behaved in precisely the opposite way to the one predicted by the hypothesis.)

It is possible to refine the hypothesis by postulating that the propensity of national governments to advantage friends and disadvantage opponents is likely to manifest itself more clearly during periods of grant cutbacks or slow growth, when only hardship can be distributed, than during periods of grant expansion. Did governments in countries with declining (or slow-growing) real grant visit the hardship disproportionately on local governments controlled by their opponents?

Unfortunately, there were only two periods of grant decline and one of relatively slow growth among the countries and periods covered. In the most severe case, real grants declined in Denmark by nearly 25 percent between 1982 and 1986; however, as has been noted, despite a Conservative central government, grants to right-wing municipalities actually declined by a larger percentage of local tax base (4.51 percent) than did grants to left-wing authorities (4.00 percent). In Sweden, real local grants declined slightly over 1983–86, but the right-wing central government did not reduce grants to left-wing municipalities disproportionately.

Are grant policies guided by objective criteria?
The objective (or bureaucratic) criteria hypothesis specifies that grants are distributed automatically according to formal rules and fixed criteria. These criteria might include one or more of a variety of factors. Grants could be awarded, for example, on a simple per capita basis, so that all recipients would receive an amount proportional to their population. They could be distributed on the basis of population size, with larger municipalities receiving disproportionate shares on the commonly held assumption that larger cities have more serious problems. Grant distribution could be based on objective criteria designed to measure indicators of local government need such as municipal unemployment rate, or it could take into account measures of municipal resource availability, such as per

capita tax base. The criterion could also consist of the rate of change in such need or resource variables rather than simply their present levels.

Below we analyze the relationship of these various objective criteria to grant change. We do so not through examining the design of grant systems or their intent, but by examining the resulting distribution of grant change among local governments. The aim is to determine whether these results are consistent with each of the objective criteria, whether they are more consistent with the partisan politics model, or whether no pattern is evident. In effect, we infer grant design and intent from grant result.

Population size   Although there was no consistent pattern across countries in the distribution of grant change by population size, grant change in several countries (the USA, Italy and France) appeared systematically to favor smaller municipalities. Larger municipalities were favored only in Norway, where the three most populous municipalities received real grant increases averaging 44.5 percent and the smallest 352 authorities had local grant increases averaging 13.3 percent, the lowest for any of the population size groups.

In France, Italy and the USA, however, it was the smaller-sized municipalities that were the 'winners' and the larger ones that were the 'losers'. In the USA during the 1980–84 period, the 748 smallest municipalities (population 25,000–100,000) averaged real grant declines of 4.0 percent, compared with 12.3 percent for the 110 cities in the 100,000–250,000 size range. These changes may to a large extent be a mere reflection of the fact that the large cities derive a larger proportion of their revenues from grants than do smaller ones. While these differences among broad size groups were significant at the 0.05 level, the simple correlation between grant change and population (logged) was small (–0.09) and not significant.

Italian municipalities with a population less than 100,000 received small real grant increases over 1980–84 (4.3 percent for those with less than 25,000 and 2.0 percent for those with population between 25,000 and 100,000), while larger municipalities suffered real grant decline (–10.3 percent for municipalities with populations between 100,000 and 250,000, and –14.2 percent for those with populations above 250,000. These differences were significant at the 0.05 level, and, indeed, the correlation (–0.18) between grant change and population was also significant.

In France, the 75 municipalities with populations of 20,000–30,000 averaged real grant increases of 33.8 percent between 1979 and 1984, while the 19 municipalities with populations in excess of 80,000 had grant increases of 23.0 percent. However, the differences among these groups were not significant at the 0.05 level. The simple bivariate correlation for all municipalities between grant change and population (logged) was –0.17.

In Sweden, Denmark and Finland, there were no significant differences among the different population size categories in terms of grant change.

What patterns can be discerned from this apparently diverse behavior? If grant formulas are structured so as to favor larger cities because of their presumed greater need, if they are not altered during times of economic change, and if they are symmetrical in times of both grant increases and grant decreases, then we could expect both increases and decreases to be directed disproportionately toward larger cities. Thus, the formula would work to provide relatively more funds to large cities during times of grant increases, but it would also work to take more away from larger cities during times of grant decline. This hypothesis would explain the case of Norway, where grant increases went disproportionately to large cities, and the USA, where grant decline also disproportionately disadvantaged the larger cities. However, in France and Italy grant increases were disproportionately distributed to smaller towns (although we must remember that these findings did not meet the test of significance in France and that the grant increases were very small in Italy).

Resources   If local resources are measured by per capita income, again, on first analysis no common pattern is apparent. In Denmark (at least over 1978–82), Italy and France, grant change appeared to advantage poorer communities; in Norway, Finland and Sweden wealthier communities benefited, and in the USA there was no apparent trend.

In Denmark the poorer municipalities received disproportionately high grant increases during the 1979–83 period, a period of overall national government control by the left; grant change was strongly associated (negatively) with municipal wealth, as measured by per capita income ($r = -0.56$). However, from 1982 to 1986, while there was no overall correlation between wealth and grant change, the wealthiest quartile of municipalities none the less averaged a grant decrease equal to only 4.22 percent of their tax base, compared with a 4.67 percent decline for municipalities in the poorest quartiles.

In Italy, grant increased to municipalities in the poorest quartile by 16.5 percent, while declining to municipalities in the wealthiest quartile by 8.2 percent. The correlation between grant change and wealth was -0.36 and was significant at the 0.001 level.

In France, grant change over the 1980–84 period also advantaged the poorer municipalities. Grants to municipalities in the poorest quartile increased by 73.8 percent compared with 52.4 percent for those in the wealthiest quartile. However, the entire change occurred over 1980–82; during the 1982–84 period, there was essentially no difference in grant change to wealthy and poor municipalities.

In Norway, on the other hand, wealthy communities quite clearly were advantaged by grant change over the 1981–84 period; grant was positively associated with wealth ($r = 0.23$), indicating that wealthier municipalities received higher grant increases than poorer municipalities. In Finland, too, wealthier municipalities appear to have made out slightly better than poorer ones over the 1980–84 period. While the overall correlation between wealth and grant change was slight (0.13) and not significant at the 0.01 level, the wealthiest quartile of Finnish municipalities averaged grant increases of 44.4 percent while each of the other three quartiles averaged increases of only 37–38 percent. In Sweden, there was a slight (and statistically significant) correlation between grant change and wealth over 1983–86 ($r = 0.15$), but no correlation over the longer 1978–86 period. In the USA, there was no correlation overall between wealth and grant change from 1980–84 ($r = 0.04$).

Can either the partisan politics or the objective criteria hypothesis impose any order on this seemingly diverse experience? The partisan politics hypothesis would suggest that national governments controlled by left parties would provide grant increases disproportionately to poorer communities (where their supporters are more likely to reside), while national governments controlled by the right would give grant increases disproportionately to wealthier communities. The objective criteria hypothesis would suggest grant increases would go disproportionately to poorer communities, while grant decreases, unless formulas are changed or are not symmetrical, would be directed disproportionately to the same areas.

The data do seem to point in the direction of the partisan politics hypothesis. In Denmark, the left government in office during the earlier period sent grants disproportionately to poorer areas, while the right government that followed imposed grant reductions disproportionately on the same areas. However, the above argument is completely false in the Danish case, as the Agrarian Party (Venstre), which is a member of the right-wing coalition government, has its strongholds in the small, rural and poorest areas of the country. (Close to half of the 275 mayorships actually were captured by the Agrarians in the 1981 local elections.)

In Norway (more so) and Sweden (less so) right-wing governments directed grant increases more toward wealthier municipalities, while in France the left-wing government sent grant increases disproportionately to poorer communities. In Italy, the coalition government clearly favored poorer communities, while in Finland, the coalition government showed a slight tendency to favor wealthier ones. In the USA, the right-wing government (perhaps functioning more as a coalition, given Democratic control of the House of Representatives) did not impose grant declines disproportionately on rich or poor areas.

The objective criteria hypothesis did not fare as well. The Danish experience is consistent with it (disproportionate gains by poorer munici-palities during a period of grant increases, and losses during a period of grant decline), as is the French experience (disproportionate gains by poorer municipalities during periods of grant increase). However, in

Norway, Sweden and Finland, wealthier municipalities gained disproportionate shares of grant increases.

Resource change In Norway, Denmark and Sweden, municipal governments experiencing favorable change in local resources (i.e. high
increases in per capita income) had disproportionately low increases in
grants. The correlation between change in per capita income and grant
change was –0.32 in Norway; in Denmark it was –0.31 for the 1979–83
period and –0.32 for the 1983–85 period (–0.38 in 1982–86); in Sweden
it was –0.43 for the 1983–86 period (although no relationship existed for
the entire 1978–86 period). In all three countries it appears that central
governments were directing grant increases more toward local governments with slowly growing tax bases, thus somewhat compensating for
the inability of these governments to raise easily increased revenues
from their own sources. This supports the objective criteria hypothesis.
However, in the USA and Finland there was no relationship between
changes in a municipality's economic base (as measured by per capita
income) and grant change ($r = 0.02$ and 0.03 respectively).

Need If need is used as a criteria for grant distribution, then the objective
criteria hypothesis suggests that grant increases should be directed disproportionately to municipalities characterized by high levels of economic
distress. However, in only two of the countries was distress of municipal
economies (as measured by local unemployment rates) highly and significantly related to grant change. In Italy, the correlation between grant
change and municipal unemployment rate was 0.47: municipalities with
higher unemployment rates received higher grant increases. In Sweden,
however, grant change was inversely related to high unemployment rates
over 1978–86 ($r = –0.33$); thus, municipalities with higher unemployment
rates received disproportionately lower grant increases. However, there
was no correlation during the 1983–86 period. In Denmark there was a
positive correlation over 1982–86 of 0.20, indicating a slight tendency
for municipalities with high unemployment rates to receive greater grant
increases; in France and the USA there were negative correlations of
approximately –0.10 and in Norway of 0.08.

   If we assume that municipalities with high unemployment rates have a
high proportion of residents supporting left parties, while municipalities
with low unemployment rates are filled with supporters of right parties,
then the partisan politics hypothesis suggests that left-wing national
governments should direct grant increases disproportionately to areas
of high unemployment rates and right-wing governments to areas of
low unemployment rates. Indeed, in Sweden the right-wing government
during the 1978–82 period did direct grant increases disproportionately to
areas of low unemployment rates. However, in Denmark the right-wing
government did exactly the opposite during the 1982–86 period.

REPLACEMENT BEHAVIOR

What was the response of local governments to loss of grant? To what extent did they replace lost grant with increased own-source revenue? Over the entire 1979–84 period, local government systems in five countries experienced declines in real grant (see Table 5.4). In three of these countries (Italy, the USA and Canada) increases in own-source revenues by local governments more than made up for grant losses. In Italy, increases in own-source local revenue in 1979–84 exceeded grant loss over the same period by 42 percent, in the USA by 13 percent and in Canada by 12 percent. In the other two cases (Germany and the UK) increases in own-source local revenues partially compensated for grant declines: in Germany own-source local revenue increases replaced 84 percent of lost grant, while in the UK the corresponding figure was 49 percent.

During the slow-growth 1979–82 period, replacement behavior was less impressive. German local governments replaced none of their lost grant and, indeed, suffered additional losses of own-source revenue, while US local governments replaced only 8 percent of lost revenues, UK local governments 41 percent and Canadian local governments 57

Table 5.4   *Replacement behavior*

	1979–82			1982–84			1979–84		
	A	B	C	A	B	C	A	B	C
Denmark				−1499	1667	111			
Norway									
Sweden									
Finland									
Germany	1134	−396	−35	−1711	2797	163	−2845	2401	84
UK	2002	814	41				−1411	688	49
France									
Italy	202	204	101	−233	414	178	−435	617	142
Canada	244	140	57	−87	231	266	−331	371	112
USA	2676	217	8				−2364	2681	113

A: decline in real grant.
B: change in real own-source actual revenue.
C: percentage replacement = $(B/A) \times 100$.

The replacement value is only calculated for countries that experienced declining real grants. Figures in the first two columns for each period are in the national currencies (in millions). Thus, 1667 for Denmark means DKr 1,667,000,000.

Unit of analysis: countries.

percent. Only in Italy did local governments replace all of their lost grant with own-source revenue. By contrast, in the more robust economy of 1982–84, local government systems in all four of the countries where real grants declined (Denmark, Germany, Italy and Canada) replaced all of the lost grant with additional own-source revenue.

Data on replacement behavior in individual cities are available only for the USA for 1980–84 and for Denmark for 1982–86. Replacement behavior was clearly related to fiscal stress. In the USA the most fiscally stressed cities (scores below 90) replaced none of their lost grant in real terms, and, indeed suffered additional losses in own-source revenue. Moderately stressed cities (scores 90–100) that lost grant, however, raised sufficient additional own-source revenue to completely replace lost grant; indeed, additional revenue exceeded lost grant by an average of 49 percent.

In Denmark even the most fiscally stressed set of cities (fiscal *SLACK* scores below 80) more than replaced their entire grant loss, and the less fiscally stressed a city was, the more it 'overreplaced' its lost grant. Indeed, the simple correlation between fiscal *SLACK* score and grant replacement was 0.60, and in a regression model explaining replacement, the fiscal *SLACK* score was significant and had the highest beta of any of the variables (change in population, change in unemployment, per capita income, tax rate and party control).

It seems clear that replacement behavior is, in general, much easier during times of economic growth than during times of stagnation or decline. When local tax bases and thus tax revenues expand with economic growth, politicians can replace lost grant with increased local spending without taking the politically visible step of raising local tax rates. Instead, tax revenues expand automatically. However, tax bases do not always automatically expand with economic growth. In England, where the rates (local property tax) comprised the only permitted local tax before 1990, local property rateable values had not been reassessed (a central government responsibility) since the early 1970s; thus, even during periods of economic growth local authorities must increase the tax rate in order to raise additional local revenues. This fact undoubtedly helps to explain the very low replacement ratio in the UK (only 49 percent, by far the lowest of any country) between 1979 and 1984.

GRANT POLICIES AND LOCAL GOVERNMENTAL SYSTEMS

It is symptomatic of all the countries investigated that central government policies toward local government – and this is particularly true for grant policies – are created on an ad hoc basis with a high degree of trial-and-error experimentation involved. No grand long-term master plan was guiding the conservative governments in Denmark or the UK during the first years of belt-tightening. The extent and distribution of cutbacks, the use of limits on taxation, spending or borrowing and the use of sanctions shifted from year to year as a function of how effective the various instruments were perceived to be, what the political costs turned out to be, and how the national economy and central government fiscal health appeared to develop.

In such an atmosphere of policy-making, it is highly unlikely that anyone, whether in central government, local government associations or individual local governments, evaluates the long-term effects of this volatile and ever shifting mix of instruments, resource flows and punishment systems.

Such an evaluation is the central concern of the following section. Admittedly, the time span is too short for a thorough evaluation of the impacts of central government policies on the working conditions of local governments. Nevertheless, changes were quite dramatic in the period under study and ought to have some consequences for the four aspects we intend to investigate:

1 Have local governments become more or less vulnerable?
2 Has local autonomy, defined here as fiscal discretion, increased or decreased?
3 Is equalization greater or smaller?
4 Have local disparities increased or decreased?

Vulnerability
After three years of cutbacks in general grants to local governments, the chairman of the Danish Association of County Authorities had had enough. Fiscal planning for more than one year seemed impossible because government policies were totally unpredictable. In 1985 he therefore suggested that general grants be abolished, leaving the counties to raise all revenues from local tax sources.

The chairman of the Association was clearly in line with many local political leaders at the time. Grants are a mixed blessing. They provide resources and often act as a shield, partially protecting local governments from downturns in the local economy, but also making them more vulnerable and dependent upon the uncontrollable and often unpredictable policies of central government. The commonly accepted measure of dependency or vulnerability is the amount of intergovernmental grants as a percentage of total local government revenue.

Using this measure, the local government systems in Italy, Canada and the UK were all relatively highly 'vulnerable' at both the beginning of the period (1979) and the end (1985); cf. Table 5.5. In each case grants constituted more than 40 percent of total revenue in both time periods (and, in Italy, more than 50 percent). Local government systems in

Table 5.5 *Vulnerability*

| | Grants as % of total revenue | | |
	1979	1982	1985
Denmark	33.4	35.6	28.0
Norway	29.4	30.7	32.4
Sweden	26.0	25.6	25.4
Finland	18.2	18.5	20.7
Germany	31.3	30.1	28.1
UK	46.8	41.3	43.4
France	34.9	35.9	34.7
Italy	59.4	56.2	52.5[1]
Canada	50.9	49.4	47.7
USA	37.0	33.1	30.4
Mean	36.7	35.7	34.3

[1] 1984.
Unit of analysis: countries.

Finland, Sweden and Norway all had the lowest degree of vulnerability (less than 30 percent) in 1979, while Finland, Sweden and Denmark had the lowest vulnerability in 1985.

How has the change in national government grant behavior, when combined with local government's own fiscal behavior, altered local government 'vulnerability'? On average, grant dependence declined slightly for the 10 countries, from 36.7 percent in 1979 to 35.7 percent in 1982 and to 34.3 percent in 1985. This included substantial changes over the 1979–85 period in two countries: from 33.4 to 28.0 percent in Denmark, and from 37.0 to 31.3 percent in the USA. In addition, grant as a percentage of total revenue declined in Italy from 59.4 percent in 1979 to 52.5 percent in 1984 (1985 data were not available.) In all three countries this reflected substantial real declines in grant rather than an expansion of locally raised revenues. Of the three countries with the highest vulnerability scores, Italy (as noted) registered a substantial decline and Canada and the UK more modest declines.

Grant dependence actually increased slightly over the 1979–85 period (from 28.5 to 29.0 percent) for local government systems in countries that had local government fiscal *SLACK* scores above 100 during this period, and declined (from 42.3 to 37.9 percent) in those countries in which local government systems experienced fiscal stress (had fiscal *SLACK* scores below 100). Since fiscal stress results primarily from lost grant, this simply means that, having lost grant, local government systems in these countries are now less vulnerable than previously.

It was hypothesized that during periods of grant reduction countries whose local government systems are highly vulnerable (that is, whose

Table 5.6   *Vulnerability and fiscal stress of countries with grant reductions*

	Vulnerability score, 1982	Fiscal slack score, 1979–85
Italy	52.5	95.2
Canada	47.7	96.0
UK	43.4	91.0
Denmark	35.7	99.3
USA	30.4	98.3
Germany	28.1	101.1

Unit of analysis: countries.

local governments derive a high proportion of their revenue from grant) also would have local government systems experiencing a greater degree of fiscal stress (low fiscal *SLACK* scores); cf. proposition 2c above. That hypothesis appears to be confirmed. During the 1979–85 period, 6 of the 10 countries experienced real grant declines. The local government vulnerability score at the midpoint year (1982) and the local government fiscal *SLACK* score over the 1979–85 period are given in Table 5.6 ranked by vulnerability score ($r^2 = 0.49$).

The three countries whose local government systems had the highest degree of grant dependence (Italy, Canada and the UK) also had the most fiscal stress. While the UK ranked only third in vulnerability, it had much larger grant reductions than did the other two countries (15.1 percent compared with 4.5 percent in Italy and 5.2 percent in Canada), thus explaining why it had the highest degree of fiscal stress.

While grant systems, depending on the extent and mechanisms of equalization, may protect individual local governments from a declining local resource base, a large degree of dependency, on the other hand, makes local governments more vulnerable to downturns in the national economy. The fiscal crisis of the state is more likely to be exported to local governments the more dependent these are on monies coming from central government. This finding is particularly interesting in relation to the findings made earlier about the nature of the grant system: dependency seems to be a more important condition than the nature of the grant system (consolidated or fragmented) when it comes to understanding the extent to which the fiscal crisis of the state is exported to local governments.

Local autonomy (fiscal discretion)
Change in 'vulnerability', as defined above, does not necessarily translate into change in local autonomy. Conceptually, local autonomy relates to the degree of discretion in local government. If local government is highly dependent upon conditional grants it will have little discretion.

Table 5.7   *Local government budgetary discretion*

	Discretion score			
	1979	1982	1984	1985
Denmark	0.77	0.77	0.78	0.77
Norway	0.75	0.74	0.71	0.72
Sweden	0.73	0.75	0.75	0.75
UK	0.81	0.84	0.77	0.77
France	0.81	0.81	0.82	–
Canada	0.56	0.54	0.55	–
USA	0.70	0.72	0.73	0.74
Mean	0.73	0.74	0.73	

Budgetary discretion = Total taxes + Fees and charges + General grants/Numerator + Discretionary grants.
Data were not available for Finland, Germany and Italy.
Unit of analysis: countries.

Operationally, then, local autonomy can be measured by discretionary revenues (taxes, fees and user charges and general grants) as a percentage of total revenues (discretionary revenues plus conditional grants). As Table 5.7 indicates, the UK, France and Denmark had the highest degree of fiscal discretion in 1979 and 1984–85, while Canada and the USA had the lowest. There was very little change over the 1979–84 and 1985 period for most of the seven countries for which data on discretion are available. The largest change occurred in the UK where the degree of local government fiscal discretion rose from 81 percent in 1979 to 84 percent in 1982 and then fell to 77 percent in 1985. Fiscal discretion also fell slightly in Norway, from 75 to 72 percent between 1979 and 1985. The degree of fiscal discretion increased slightly in the USA (70 to 74 percent). There was no discernible relationship between fiscal stress and change in fiscal discretion.

Equalization
In the discussion on resources above, we examined whether changes in grants over the period studied were directed disproportionately toward poorer communities. We now turn our attention to the entire grant system at a single point in time, rather than to changes in grants over time. Grant systems frequently serve to compensate for disparities in resources among local governments by providing disproportionate shares of grant to poorer communities. To the extent that they do so, they have an 'equalizing' effect on sub-national disparities. Three of the four Nordic countries have the most equalizing grant systems. The correlation between grant per capita and income per capita was –0.83 for Denmark (1986), –0.58 for

Table 5.8   *Equalization effect of grant system (Pearson's* r)

	Early period	Late period
Denmark	–0.64 (1982)	–0.83 (1986)
Norway	–0.37 (1981)	–0.45 (1984)
Sweden	–0.08 (1983)	–0.07 (1986)
Finland	–0.54 (1981)	–0.58 (1984)
France		0.04 (1984)
USA	–0.23 (1980)	–0.18 (1984)

The figures show the correlation between grant per capita and income per capita.
Unit of analysis: municipalities.

Finland (1984) and –0.45 for Norway (1984). However, Sweden's system had almost no equalization impact (a correlation of –0.07 in 1986) and France also had virtually none (0.04 in 1984). The US grant system had a very modest equalization effect (–0.18 in 1984) (see Table 5.8).

In both Denmark and Norway the equalizing impact of the grant system increased over the course of the period studied. In Denmark the correlation between grant per capita and income per capita increased from –0.64 in 1982 to –0.83 in 1986. This increase reflected major changes that took place in the way Danish grants were distributed. The new Conservative government, which took office in 1982, imposed grant reductions that were first aimed to hit wealthy municipalities the hardest; then, after the initial cutbacks were imposed, the reductions were allocated according to change in municipal tax bases, with municipalities with fast-growing tax bases receiving less grant.

In Norway, the correlation between grant per capita and income per capita increased from –0.37 in 1981 to –0.45 in 1984, while in Finland the equalization impact increased slightly, from –0.54 in 1981 to –0.58 in 1984.

In the USA, on the other hand, the relatively small equalization tendency of the grant system actually diminished, declining from a correlation of –0.23 in 1980 to –0.18 in 1984. In Sweden, where the grant system appears to play virtually no equalization role, the correlation between grant per capita and income per capita remained virtually unchanged (–0.08 in 1983 and –0.07 in 1986).

Concentration/dispersal
Local government systems can be considered concentrated if local governments within the system behave in a relatively homogeneous manner and dispersed if there is substantial variation in their behavior. Dispersion is frequently viewed as more conducive to innovation.

A high degree of concentration may result from the grant system's

Table 5.9   *Concentration/dispersal of local government systems*[1]

	Early period	Late period
Denmark	0.114 (1979)	0.119 (1985)
Norway	0.255 (1981)	0.287 (1984)
Sweden	0.177 (1982)	0.181 (1985)
Finland	0.211 (1981)	0.188 (1985)
France	0.271 (1979)	0.243 (1984)
USA	0.643 (1980)	0.661 (1985)

[1] Measured by coefficient of variation for per capita expenditure.
Unit of analysis: municipalities.

ability to equalize opportunities between local governments. It may, however, also indicate the presence of stronger national service standards. Finally, a high degree of concentration could come about as a result of homogeneity across jurisdictions with respect to local economic resources or political culture.

Current expenditure per capita is taken as a measure of fiscal behavior, a behavior which, as has been shown, is affected by changes in government grant. Concentration/dispersal is measured by the coefficient of variation of per capita expenditure (standard deviation divided by the mean). Equalizing grant systems (see above) reduce the level of dispersion that would otherwise occur in a local government system.

Of the six countries for which data are available, Denmark had the most concentrated local government system, while the USA had the most dispersed (see Table 5.9). The latter clearly is an outlier, as the coefficient of variation is more than twice as large as in any of the other countries (reflecting not only a low degree of equalization but also the fact that functions vary across US local governments much more than in the other countries).

Denmark's highly concentrated system may in part reflect its highly equalizing grant system described in the previous section. If the equalization schemes were abolished altogether, the coefficient of variation would be twice as large, an indication of a degree of equalization in the Danish system of about 50 percent.

In Norway, local government system behavior becomes somewhat more dispersed over the time period examined. The coefficient of variation for current expenditure per capita increased by 12.5 percent between 1981 and 1984, from 0.255 to 0.287. In both France and Finland, local government behavior became more homogeneous. In Finland the coefficient of variation fell from 0.211 in 1981 to 0.188 in 1985, a decline of 11.0 percent, while in France it fell from 0.271 in 1979 to 0.243 in 1984 (−10.4 percent). The decline in the diversity of local government behavior in France is particularly surprising, since it coincided with a major program of decentralization in central–local relations which, according to conventional wisdom, should have resulted in greater diversity in local government fiscal behavior. In Sweden, the degree of diversity of local government behavior remained virtually unchanged from 1982 (0.177) to 1985 (0.181). The same is the case for the USA.

SUMMARY AND CONCLUSIONS

This chapter has focused on grant change, which, as was demonstrated in Chapter 4, was the primary cause of urban fiscal stress during the period studied. It examined the causes of grant change, its distributional effects and its impact on central–local relations.

Change in GDP was found to be highly correlated with grant change (in a simple bivariate relationship) during times of slow national economic growth or decline, but not during times of relatively rapid economic growth. Similarly, during slow-growth periods governments of the right are more likely to cut grant than governments of the left.

At the very first level of analysis, there were few clear patterns in the distribution of either grant increase or grant decrease to municipalities classified by population size, local resources, change in local resources or needs. Thus, grant change did not uniformly advantage or disadvantage large or small municipalities, poor or wealthy municipalities, growing or declining municipalities, etc.

We tested two hypotheses with respect to the distribution of grant change. The partisan politics hypothesis specified that national governments would distribute grant change so that it would benefit local governments controlled by the same party or local governments with a high proportion of that party's adherents among its residents. The objective criteria hypothesis specified that grant change would benefit municipalities with high needs and/or low resources during times of grant increase, but would disadvantage them during times of grant decline.

Neither hypothesis appeared to hold across the full range of classifications used to categorize municipalities, although each did have some support in one or more of the classifications. There was no uniform trend across the countries for national governments to advantage local governments controlled by the same party and to penalize opponents. However, there appeared to be some tendency for left-wing governments at the national level to disproportionately advantage poorer municipalities in distributing grant change and for right-wing national governments to disproportionately advantage wealthier municipalities. The objective criteria hypothesis could not be discarded in analyzing the distribution of grant change by population size and by change in local resources. Clearly, the number of observations was too small and the exceptions too numerous to conclude much more than that both hypotheses deserve further analysis.

There was strong evidence that the extent of grant replacement – the replacement of lost grant with own source revenue – varied substantially with the economic cycle. During periods of slow economic growth or decline, there was relatively little replacement, while during periods of more rapid economic growth, full replacement and over-replacement were more common.

Finally, our analysis indicated that local government fiscal vulnerability (as measured by grant as a percentage of total local revenue) decreased over the period examined. More importantly, there appeared to be a strong tendency for countries whose local government systems had a high degree of vulnerability to be the same countries whose local government systems had the highest degree of fiscal stress during periods of grant reductions.

As a general pattern, we found socio-economic factors, that is, trends in the national economy and the degree of fiscal slack, to overwhelmingly dominate party politics when it comes to the content and consequences of grant policies. Our detailed analyses were not successful in isolating system-type specific patterns. (Cf. the distinction between the Scandinavian, southern European and North American models of local government in Chapter 2.) Some countries probably did group together more often than others; Finland and Norway are examples of this. In most cases, however, we explained this by reference to the existence of the same socio-economic conditions in the two countries rather than to systemic features.

The fact that few clear and undisputed generalizations emerged from the analysis should not be very surprising. In many cases, data were available for only a subset of the 10 countries and for only one or two time periods. Furthermore, each country's grant system has its own quite unique characteristics. Additional work is clearly needed to examine countries and their grant systems more intensively for commonalities and differences which can be used as variables in explaining grant behavior. What appears unique upon first analysis may well yield to the search for generalizations upon some more fundamental level of analysis.

NOTES

1 Several attempts were made to test the partisan hypothesis in a more stringent manner using the pooled data set. A so-called 'constant coefficient model' was estimated using change in GDP, party color, central government budget deficit (as a percentage of total expenditures, lagged two years) and balance of payment deficit (as a percentage of GDP, lagged one year). This model, however, produced residuals with quite different variances across countries. One possible solution to this problem is to introduce dummies for the individual countries. This so-called 'least squares dummy variable model'

exhibited an extremely high degree of multicollinearity. The data therefore do simply not lend themselves to the more stringent test, and we cannot isolate possible country effects from the more general relations, for instance between change in GDP and change in grant. For the general problem involved in analysis of pooled data, cf. Sayers (1989).

2 The relationship found was in the expected direction but it was statistically insignificant (at the 5 percent level). However, as pointed out by the authors, there remains some doubts about the validity of significance tests in non-sample data (Page et al., 1990: 49). Taking this into consideration, one can maintain that the authors found weak support for the partisan politics hypothesis. It is interesting to note, that the partisan effect seemed much stronger during the Labour years of 1976–80 as judged by the effect coefficient.

REFERENCES

Nevers, Jean-Yves 1988. 'Grant Allocation to French Cities: The Role of Political Processes', paper presented at the Annual Meeting of the American Political Science Association, Washington DC.

Page, Edward C. with Michael Goldsmith and Pernille Kousgaard 1990. 'Time, Parties and Budgetary Change: Fiscal Decisions in English Cities, 1974–88', *British Journal of Political Science*, 20: 43–61.

Sayers, Lois W. 1989. *Pooled Time Series Analysis*, Sage Quantitative Applications in the Social Sciences series, London: Sage.

PART III
LEADERS AND POLITICAL ENVIRONMENTS

6
Fiscal Stress and Local Political Environments

Harald Baldersheim

ARE FISCALLY STRESSED CITIES DIFFERENT?

The purpose of this chapter is to investigate the extent to which fiscal stress affects patterns of local leadership and the political environments of local leaders. Three aspects of local leadership are focused on: (1) city managers' 'policy maps', that is, their perceptions of problems and challenges faced by their municipalities (Eulau, 1971); (2) the mayors' efficacy or influence on local decision-making, that is, their successes or failures in implementing their policy preferences; and (3) the mayors' leadership styles, that is, their inclination to take policy positions in conformity with or against the opinion of the majority of voters.

There are plenty of stories and case studies that portray situations where fiscal stress seems to create environmental turbulence and new patterns of political leadership on the local political stage. A famous example is the fiscal crisis of New York City in the early 1970s. From France, we have among others the situation in the city of Nimes, where in the mid 1980s enraged city employees locked the top staff in their offices over a weekend after a new mayor introduced privatization of street cleaning as a response to declining city finances (Clark et al., 1987).

Even traditionally calm Scandinavian city politics have been disrupted in situations of severe fiscal stress. The city of Oslo may serve as an example in this respect. In June 1988, Oslo had its first experience of real cutbacks since World War II. The city's finances had been ailing for nearly a decade. The core of the problem was an inability to adjust levels of spending to available resources. The result was an accumulating budget deficit of nearly NKr1 billion (US$150 million) in 1988 in a city of 450,000 inhabitants. The former chief administrative officer had admonished the city council again and again to set more realistic spending levels.

The new cabinet-style city government introduced in 1986 had at first proved incapable of dealing with this problem. The local election of 1987 had brought a new coalition of right-wing populists and conservatives to power as from January 1988. Together, the new coalition managed to pass Oslo's first real cutback budget (and to vote down the Social Democratic opposition's alternative budget, based on a combination of increased property taxes and more moderate cutbacks). The budget was passed amidst intense media coverage, demonstrations staged by the city employees' unions, and the most violent street riots seen since the 1960s Vietnam demonstrations. Deputations handing in protests from nearly 20 different groups and organizations were received by the city council before the budget meeting.

Perhaps the most significant aspect of the Oslo drama was the much more visible role played by the new political leadership, the 'city cabinet' (Baldersheim and Strand, 1988). The traditional pattern of budgeting in Oslo was that cutbacks would be proposed by the chief administrative officer; the political leaders then, predictably, would react by distancing themselves from the officer's proposals and seeking to restore spending levels in the ensuing political processes. Now, with the new cabinet government, there was no chief administrator to hide behind: the cabinet of political leaders had to take full responsibility for balancing the budget. The new coalition then took an initiative to reduce expenditures unprecedented for the political leadership in Oslo, an initiative that was subsequently endorsed by a majority of the council.

The policy changes that finally took place in Oslo depended clearly on the new institutional framework that had been introduced in 1986, the so-called 'cabinet' or parliamentary system of municipal government, a system that is unique to Oslo and which represented a departure from the traditional aldermanic system that characterized Norwegian local government in Oslo. The new system allowed more clear-cut coalitions to emerge and made a clearer distinction between 'ins' and 'outs'. Media reporting was also among the forces that shaped Oslo's policy response to fiscal stress, at least marginally. One newspaper headline ran, 'The politicians force me to whore again!', a statement attributed to a former prostitute who was currently being helped by an anti-prostitution project threatened by the suggested cuts. (After that, the project was put back into the budget.)

The adoption of new policies demonstrated the inadequacy, in Oslo, of the traditional managerial approach (a learning or 'intelligence' strategy) to policy change in a political context. The city manager's strategy had been simply to present the stark figures of the city's fiscal situation, in the 'Socratic hope' (Czarniawska-Jorges, 1988) that the possession of the right kind of information would persuade the council to adopt the appropriate policies (Baldersheim, 1982, 1986). Instead, the political leaders preferred to lobby the central government for more grants rather than introduce austerity policies locally, while the central government remained unimpressed.

The new leadership pattern of Oslo seemed to reduce the influence of organized interests. Pressure group resistance had sprung up at every attempt to reduce Oslo's budget on previous occasions (Brofoss et al., 1975). This time pressure group activity was particularly intense, especially from the municipal employees' unions. At the same time, bargaining with the unions behind the scenes, which had been characteristic of previous cutback attempts, did not take place.

However, the policy changes in Oslo may also support the view that political parties and election outcomes matter. The new, radical Oslo budget was passed less than six months after an election had increased the number of right-wing populists not only in the city council of Oslo, but also in most of the other larger cities. The budget was passed with their help, and a proposal of 'no confidence' put forward by the Social Democrats was subsequently defeated with their backing.

Despite the dramatic consequences of fiscal stress described in studies of individual cities, there are other studies that suggest that many cities may be very slow to react to an adverse financial situation. Many cities seem to hold on to their policies and practices for a long time after the first signs of fiscal stress are evident. Continuity or entrenchment rather than change and retrenchment have been found to be characteristic of cities facing financial difficulties (Brunsson, 1983; Midwinter, 1988). If the latter reaction is widespread, we would not expect fiscally stressed cities to differ markedly from non-stressed cities with regard to political processes or administrative practices.

In the case of Oslo, it was quite evident that fiscal stress led to more turbulent political environments and to changes in institutional settings and leadership patterns. For the founding fathers of the new city charter, the fiscal crisis became a political opportunity to increase political influence on decision-making and to set new priorities, although considerations about the leaders' media image shaped some of those priorities. However, it took almost a decade to change the policy maps of city leaders – their perceptions of their environments – enough for them to accept that there *was* a fiscal crisis.

The general questions that emerge from this account are:

- What characterizes the policy maps of city leaders, and to what extent are they shaped by fiscal crises?
- How much influence do political leaders have on city policies?
- Is their influence diminished or enhanced by fiscal crises?
- How independent are political leaders with respect to leadership styles?
- Under what circumstances are they willing to take stands against the opinions of voters?

The data with which to address these questions come from FAUI surveys in France, Norway and the USA, which are the countries that will be examined in this chapter.[1] These countries represent three different systems of local government: the south European, north European/Scandinavian and North American models that are described by Michael Goldsmith in Chapter 2 above.

DIFFERENT SYSTEMS OF LOCAL GOVERNMENT: DIFFERENT LEADERSHIP PATTERNS?

Based on the descriptions in Chapter 2 and other sources (Page and Goldsmith, 1987; Dente and Kjellberg, 1988), some system characteristics and prevalent leadership patterns are condensed in Figure 6.1. French mayors may be said to be under the tutelage of the centre, but they are also, at the same time, advocates or representatives of their local governments at the centre, often helped by a *cumul des mandats*; decision-making seems to take place in a more intense partisan environment than in the other two countries. Norwegian mayors may be characterized as welfare state partners operating in a highly bureaucratized and standardized environment; decision-making is consensus-oriented and integrative (Larsen, 1990), with an emphasis on distributional justice, while the role of the mayor is that of an administrative 'overseer' (Goldsmith, 1987) of service provision. Relationships between different levels of government are weaker and the roles of higher levels of government less interventionist in the USA than in Europe, and so municipalities have more autonomy but also less assistance from higher up, especially from

	Central/local Relations	Decision-making Climate	Leadership Style
France	Tutelage	Partisan	Representational
Norway	Partnerships	Consensual	Administrative
USA	Autonomy	Group-oriented	Personal–entrepreneurial

Figure 6.1   *Local government system characteristics and local leadership patterns in France, Norway and the USA*

the federal level. Local party-political apparatus is weak or non-existent, so groups matter more, particularly the business community. Electoral platforms and policy coalitions are brought together more on an ad hoc basis; and personalities and personal qualities often seem to matter more as leadership qualifications than they do in Europe (Kotter and Lawrence, 1974).

These system characteristics should be seen as a summary of some dominant tendencies, not as absolute system differences. For example, political parties are, of course, not unimportant in all US municipalities, but on the other hand they are not as uniformly important as in many European countries. And while tutelage is not completely absent from Norwegian central–local relations, it is not as pervasive as in France.

The generalizations in the diagram are meant to form the basis for the formulation of hypotheses about the effects of contextual, political, organizational and personal factors on the three dependent variables: policy maps, political efficacy, and leadership style.

Hypothesis 1   City managers will be the first municipal leaders to detect signs of fiscal problems. The more fiscal stress a municipality is exposed to, the more will the manager's *policy maps* reflect a high level of crisis. Active pressure groups will increase the sense of crisis. French managers will be most inclined to attribute municipal problems to central government sources; US managers will do this least frequently, since their municipalities are less integrated with and less dependent upon higher levels of government.

Hypothesis 2   Fiscal stress will tend to reduce the *mayor's efficacy*, since there will be fewer resources to distribute and therefore fewer possibilities of maintaining coalitions. American mayors will be most vulnerable to fiscal stress, since they can blame central government policies to a lesser extent for adversity, and since they are also more dependent on free resources with which to maintain ad hoc coalitions. (See Chapter 3 on political and financial capital.) In the US case, personal qualities will also be significant for the mayors' efficacy. The efficacy of Norwegian mayors will depend more than in the other cases on the competence of their managers and their ability to control the municipal organization. Increased activity from organized groups will also increase the competition for control over municipal resources and should therefore also lead to greater difficulties for the mayor trying to implement his policy preferences. It will be easier for French and Norwegian mayors to withstand pressure groups since they are more subject to central government control and party discipline than American mayors.

Hypothesis 3   Fiscal stress will increase the need for *policy independence*; that is, mayors will have to take policy positions unpopular with the voters when resources decline. The greater policy autonomy municipalities have, the less readily can they pass the buck on to higher levels of government, and the more unavoidable such leadership positions become. The higher the activity levels of pressure groups, the more likely is it that the mayors will have to disappoint some or many of them. Policy independence is more difficult in a person-oriented polity than in a partisan one, since personalization of politics means that the mayor gets more attention and more of the blame. It is also more difficult in small communities than in larger ones, since large communities are more politically heterogeneous and therefore give the mayor more room for maneuver (Clark and Ferguson, 1983).

COMPARING PATTERNS OF POLITICS

Managers' policy maps
City managers' positions in the municipal information streams are likely to enhance their awareness of financial problems. However, playing Cassandra is another role requirement of city managers, and they may have an inherent tendency to overdo the crying of warnings against the wolf (Brunsson, 1983), with the result that other actors may stop paying attention to their warnings. If managers are the cities' early warning system, their policy maps should closely reflect levels of fiscal stress and should also be guides to the behavior of other actors in the municipalities, especially the mayors. If, however, they tend to fall into the Cassandra trap, there may be little relation between objective measures of fiscal stress and managers' warning signals about the challenges ahead. There may also be little relation between the dangers managers proclaim and the measures they themselves implement in their organizations. The way leaders talk does not always correspond to the way they act (March, 1984; Brunsson, 1985).

The managers' policy maps have been constructed from responses to questions put to the chief administrative officers about the sources of their cities' financial problems. The administrative officers were asked to indicate how important a number of possible problem sources were with regard to the financial situation of their respective cities. The responses have been summarized as a 'problem index'. The problem sources ranged from 'new tasks imposed by central/regional/state governments' to local factors such as unemployment or rising service demands from citizens.

In Table 6.1 the problem sources for each country are listed. The scores are averages for municipalities in each country, on a scale from 0 to 10; the same scale will be used for most other variables; however, comparison of the absolute scores across countries should be treated with caution, although the rank order of items can be compared. (See the

Table 6.1   *Managers' policy maps: sources of financial problems (scale 0–10; 10 highest)*

Q: 'In the last three years, how important have each of the following problems been for your city's finances?'[1]

Norway	France	USA
1 Mandated costs from central govt 5.95	1 Central govt restriction on fees and charges 6.52	1 Inflation 6.67
2 Loss of tax revenues 5.13	2 Level of central govt grant for current expenditures 6.22	2 Loss of state revenue 5.48
3 Reductions of central government grants and subsidies 4.81	3 Problems in obtaining loans 6.20	3 Municipal employee pressures 4.67
4 Rising service demands from citizens 4.66	4 Local economic situation 6.09	4 State tax revenue or expenditure limits 4.66
5 Politicians unwilling/ unable to set clear priorities 4.30	5 Inflation 5.64	5 Federal and state mandated costs 4.36
6 Inflation 4.13	6 Unemployment 5.58	6 Loss of federal revenue 4.34
7 Unemployment 3.46	7 Level of central govt grant for investment 5.49	7 Increased service demands from citizens 4.01
8 Resistance to change in municipal administration 1.76	8 Loss of tax revenue 5.44	8 Unemployment 3.81
9 Pressures from municipal employees 1.53	9 Increased service demands from citizens 5.11	9 Local pressure to to decrease taxes 3.74
10 Central govt restrictions on expenditures or charges 1.14	10 Demographic evolution 4.33	10 Declining tax base 3.35
11 Problems in obtaining loans 1.14	11 Fiscal consequences of decentralization 4.23	11 Bond referendum failure 0.92

Table 6.1   *continued*

Norway	France	USA
12 Pressures from local taxpayers to reduce taxes and spending	12 Grants from regions	
1.05	3.86	
	13 Grants from departments	
	3.85	

[1] Corresponds to Q.1 in French questionnaire and Q.2 in Norwegian.

technical Appendix to this volume for a discussion of the standardization problem.)

According to the French chief administrative officers, the most important source of problems is central government restrictions on fees, charges and taxes; in Norway it is 'new tasks imposed by the central government', while US administrative officers are most concerned about inflation.

The three sets of municipalities seem to be faced with the consequences of rather dissimilar central government policies at the time of the surveys. The French municipalities (1985) are exposed to a restrictive central government trying to limit local activities (reduction of grants and restrictions on fees and charges). Norwegian municipalities (1986) feel under pressure from the central government to expand services but with a shrinking tax base. In the USA (1983–84) a loosening of fiscal ties between levels of government seems to occur, with a reduced responsibility especially for the federal government (the New Federalism dawns?). If expectations about assistance from higher levels of government are lower, it is understandable that US managers put more emphasis on other sources of policy problems such as inflation and salary demands from municipal employees (especially if the latter are driven by inflation)

In the French and Norwegian cases, two of the three most important sources of policy problems have to do with the central government. In the American case, only one of the three refers to higher levels of government. I have, however, tried to find a more precise measure of how much of the managers' policy problems they attribute to higher levels of government as against local factors that are closer to their own control (such as salary demands by city employees). A measure for central government problem attribution has been calculated: the score for Norway is 124, for France 112 and for the USA 109. This means that, when the central government problem attribution is taken out as a separate index and then expressed as a percentage of the total problem index, the former is 24 percent higher than the total for Norway, 12 percent higher for France and 9 percent higher for the USA. In other words, in all three countries the central government looms larger, in the minds of the city managers, as a problem generator than do more local factors or the general state of the economy, such as inflation. The city managers all see their cities as parts of a wider political system and also as victims of the system, most so in Norway and least so in the USA. This is, of course, not an unnatural view, since they are all, to some extent, dependent on grants and legislation from higher levels. However, of the three, Norway was the country where municipalities overall experienced the largest growth in grants in the years before the survey (cf. Chapters 4 and 5); still, complaints over the central government were most widespread there. It is therefore tempting to see in this pattern not only a role bias on the part of city managers, regardless of local government system, which makes them complain about the central government, but also a system difference between the three countries, where Norway seems to be the country where municipalities are more integrated with the central government than in the other two. This is rather against the traditional belief, where France is often regarded as one of the most centrally controlled countries of Western Europe.

Pressure groups

The measures of pressure group activity are based on the mayors' own reports on who the active and influential groups were. In Table 6.2, the active pressure groups have been listed for each country according to their perceived level of activity. The summary measure of pressure group activity is an additive index based on the scores of 11 to 13 different groups.

The countries differ with regard to their most active groupings: cultural associations (France), youth and sports clubs (Norway) and city employees (USA). However, there are also similarities. The two top-ranking groupings are the same in Norway and France (sports/youth and cultural associations). Trade unions and business and industry are among the top three in both Norway and the USA.

This pattern may reflect variations among these countries as to the functions of municipalities. American municipalities are more concentrated around regulatory functions and collective community services, with an emphasis on promoting economic growth and with an eye on the constant competition between cities (Peterson, 1981). Municipalities and business/industry are, therefore, natural allies. It is interesting to note that the most active groupings in France and Norway represent areas in which municipalities are active as financial supporters of voluntary organizations (sports and cultural associations) but do not produce a large volume of services themselves.

Table 6.2   *Pressure-group activity (groups listed in descending order of activity level)*

Q: 'Please indicate how active the participant has been in pursuing (his) spending preference.'[1]

Norway	France	USA
1 Citizens	1 Cultural associations	1 Public employees and their associations
8.15	7.31	5.23
2 Youth or sports associations	2 Services	2 Business and business-orientated groups
7.17	6.53	4.99
3 Cultural associations	3 Youth	3 Neighbourhood organizations
5.92	5.93	4.72
4 Trade union associations	4 Elderly	4 Elderly
4.88	5.68	4.66
5 Business and industry	5 Shopowners	5 Local media
4.85	5.68	4.33
6 Ad hoc groups	6 Neighborhood organizations	6 Citizens
4.31	5.02	4.34
7 Humanitarian organizations	7 Users' organizations	7 Organizations concerned with low-income groups and families
3.98	4.95	4.01
8 Residential association	8 Communal personnel	8 Organizations concerned with minority groups
3.77	4.66	4.21
9 The elderly	9 Labour unions	9 House-owners' organizations
3.45	4.52	4.16
10 Religious organizations	10 Chambers of commerce	10 Civic groups
2.43	4.37	4.02
11 Property owners	11 Tenants' organizations	11 Taxpayers' associations
1.49	4.20	3.49
	12 Environmental groups	12 Religious organizations
	3.74	2.57
	13 Business organizations	
	3.62	

Table 6.2   *continued*

Norway	France	USA
	14 Citizens	
	3.32	
	15 Taxpayers' associations	
	2.29	

[1] Corresponds to Q.20 in French questionnaire and Q.23 in Norwegian. The Norwegian wording is slightly different but also allows a rank ordering of groups in terms of their pressure activities ('Were you contacted one or more times by the groups or organizations mentioned below in connection with the making of the 1985 budget?').

Organizational development

The level of organizational development is measured by the municipality's score relating to the adoption of various information-gathering and processing techniques aimed at improving decision-making and administrative control (revenue forecasting, fiscal information systems, performance measures, accounting and financial reporting, management rights, measurement of economic development, methods of reporting to council). The adoption of these techniques is analyzed in Chapter 11, so they will not be presented in detail here. Only a summary measure of such organizational changes is used, and then as an independent variable that may affect, say, the mayor's efficacy.

The mayor's efficacy

The mayor's efficacy is measured by the mayor's response to a question about his success in implementing his own spending preferences in his present term of office. Again, a summary measure for political control has been constructed by adding the mayors' scores for success on a series of individual spending purposes. This summary measure is used later on in Tables 6.5 and 6.6. Here we look at the individual areas concerning the mayors' perceptions about their success (Table 6.3).

French mayors feel they have been most successful in the fields of garbage collection, education and infrastructure. Norwegians feel most successful with regard to housing projects, services for the elderly and water and sewage projects, while US mayors think they have succeeded best with police protection, controlling the number of municipal employees and fire protection.

The common denominator is that mayors in all three countries feel they have been successful with regard to collective infrastructure services

Table 6.3   *Mayor's efficacy (mean scores for service areas)*

Q: 'Please indicate how successful you have been in implementing your own spending preferences in this term of office.'[1]

	France	Norway	USA
Education	7.4	5.6	2.7
Health	6.9	5.7	4.2
Elderly		6.1	
Kindergarten		5.4	
Social aid	7.2	5.3	4.5
Roads		5.4	7.3
Fire		4.8	7.5
Transport	6.5	4.1	5.2
Housing	7.2	6.5	6.0
Parks, recreation		4.8	7.4
Youth		5.3	
Sports		5.7	
Revitalization		5.9	
Church		4.9	
Culture		5.2	
Water/sewage		5.9	
Municipal wages		4.6	7.4
Planning, financial control	6.6	5.6	
Municipal employment	7.1	5.0	7.6
Equipment, capital stock	6.7		7.3
Police, security	7.2		7.7
Infrastructure	7.3		
Garbage collection	7.7–6.7[2]		
Mean, all areas	6.8	5.4	6.6
N (max.–min.)	(66–60)	(365–267)	(330–171)

[1] Corresponds to Q.16 in French questionnaire and Q.21 in Norwegian.
[2] *Voirie.*

(garbage collection, housing, water/sewage and fire protection). Only the Europeans rank themselves as highly influential in personal services (education – France; the elderly – Norway). US mayors stand out with their emphasis on their successes in promoting police protection. This fact probably reflects American urban problems, although security is an area that French mayors also seem to give a good deal of attention. (In Norway, police services are not a municipal responsibility.)

Leadership style: paternalism or populism
The role of mayor requires the office-holder to reflect and represent public opinion but also from time to time to try to shape opinion or sell ideas to voters. Mayors have to strike a balance between a reflective and a

Table 6.4 *Leadership style: policy position contrary to dominant opinion*

Q: 'Sometimes elected officials believe that they should take policy positions which are unpopular with the majority of their constituents. About how often would you estimate that you took a position against the dominant opinion of your constituents?'[1]

	France (%)	Norway (%)	USA[2] (%)
Never	7	7	6
Rarely/from time to time	72	73	64
Quite often/once a month	19	18	21
Most of time/regularly/monthly	2	2	10
	100	100	100
N =	112	378	328

[1] Corresponds to Q.32 in French questionnaire and Q.10 in Norwegian.
[2] The US item had 5 categories; categories 4 and 5 have been combined.

proactive leadership, in other words, between populism and paternalism. Increasing fiscal stress will probably require them to become more paternalistic. They may have to tell the voters increasingly unpopular truths – about taxes or fees that have to be raised, services that have to be cancelled or promised policies that cannot be pursued.

It was suggested earlier that going against the dominant opinion among voters would be easier for mayors protected by a party or a partisan environment, mayors only indirectly elected (by the council, not directly by the voters) or mayors from larger municipalities. Paternalism is expected to be more pronounced in a culture with hierarchical political and social traditions (Clark, 1985) and also to be more frequent among mayors from parties with hierarchical traditions.

The more often mayors say they take positions they know are unpopular with the voters, the more they demonstrate a paternalistic inclination. Table 6.4 contains a comparison of mayors' behavior in this respect. The patterns of leadership styles are quite similar in the three countries. The greater proportion of mayors go against popular opinion from time to time, but not very often. Are they all populists then, or do mayors simply not respond to the strains of fiscal stress?

HOW IMPORTANT IS FISCAL STRESS FOR CITY POLITICS?

Fiscal stress
The indicator of fiscal slack was discussed and defined in Chapter 3. Municipalities have been divided into two groups, those with high scores

and low scores on the slack indicator. The cut-off point between 'high' and 'low' is the mean for each country. The mean varies, since the overall level of fiscal slack/stress varies from one country to another, with Norway having the highest level of slack of the three countries and the USA having the lowest. If, for example, the mean for the US municipalities had been used also for France and Norway, there would have been too few cases below the mean for meaningful analysis. Below the mean we speak of 'stress municipalities'; above the mean we speak of 'slack municipalities'.

Table 6.5 shows bivariate relationships between fiscal stress and the variables that have been discussed so far. The table gives group means for the two sub-groups (stress and slack cities) and for the total sample, and also the bivariate correlation coefficients between fiscal stress and the other variables. The means for the two sub-groups are, on the whole, very similar. Few significant F-values were found, with the exceptions of

Table 6.5   *Fiscal slack impact (group means, one-way analysis of variance and bivariate correlations) (Pearson's r)*

	Fiscal slack group means		Total means	Standard deviation	Correlation
	Low	High			
Policy maps					
Fr	5.1	5.1	5.1	1.2	−0.02
No	3.1	2.8	3.3	1.3	−0.10
USA	4.3	4.2	4.2	1.4	−0.09
Pressure group activity					
Fr	4.9	4.8	4.9	1.3	−0.01
No	4.7	5.0	4.9	2.3	0.06
USA	4.6	4.1	4.3[2]	1.5	−0.16[*]
Mayor's efficacy					
Fr	6.9	6.8	6.9	2.1	−0.06
No	5.3	5.4	5.4	1.1	0.001
USA	6.5	6.9	6.7[1]	1.6	0.10
Organizational development					
Fr	4.1	4.6	4.3	1.7	0.13
No	3.3	3.1	3.2	2.5	−0.07
USA	6.4	6.1	6.3	1.6	−0.10
Leadership style					
Fr	3.6	3.8	3.8	1.9	0.02
No	2.8	2.9	2.9	0.6	0.04
USA	3.4	3.5	3.4		−0.02

N: see Tables 6.1–6.4.

[1,2] One-way analysis of variance has yielded F-values significant at the 0.05 and 0.01 levels, respectively.

[*] Significant at the 0.05 level.

pressure group activity and mayors' efficacy in American municipalities.

The only significant correlation coefficient is between fiscal slack and pressure group activity in US municipalities. The coefficient is negative; that is, the less slack, the more pressure group activity, as stated in hypothesis 1. The reason why the fiscal slack variable has some impact only in the USA may be that the US cities are hit so much more by fiscal stress than French and Norwegian ones. In other words, the US cities were at this point in time at a position on the slack index where certain effects could be felt, while the Norwegian localities, for example, were sufficiently high up on the index for the actual score not to reflect any effect from pressure group activity.

The managers' evaluations of external challenges (policy maps) are not affected by objective levels of fiscal stress. Managers seem to feel that they have reason to complain in both sunshine and rain, as true Cassandras. This is a clear parallel to the discussion in Chapter 3 about the psychological and political aspects of fiscal crises. Also, managers do not resort more vigorously to organizational development in times of financial difficulties, but do it independently of such situations.

The mayors' efficacy does not seem to wax and wane with the financial fortunes of their cities, and their styles of leadership also seem unaffected by fiscal stress. They do not become more paternalistic when city money becomes more scarce.

Leadership patterns: products of political, organizational, personal or contextual variables?
Table 6.6 reports bivariate results for the other hypotheses concerning leadership patterns. I shall comment on the table in terms of the hypotheses suggested earlier, although many of the relationships are very weak.

Policy maps   The expectation that high levels of pressure group activity would increase the managers' sense of crisis is not quite without foundation, at least not in the US case.

Mayors' efficacy   Pressure group activity seems unrelated to mayors' successes or failures in implementing their policy preferences. So at least we may conclude that mayors who fail cannot blame pressure groups for their failure. Mayors of the right are more effective than mayors of the left, it seems; this is the case in Norway and the USA but not in France (where the coefficient is not significant). Sophisticated organizational development was expected to enhance the mayors' political control, especially in Norway. This does not happen in Norway (nor in the USA), but such a relationship is found for France. Personal qualities do not matter in France or the USA (where they were supposed to matter most), but education has some effect in Norway.

Table 6.6   *Local leadership and political, organizational, personal and contextual variables and bivariate correlation coefficients*

	Policy map (manager)			Mayor's efficacy			Mayor's policy independence		
	Fr	No	USA	Fr	No	USA	Fr	No	USA
Political									
Party affiliation (right/left)	0.17	0.04	0.12	0.10	–0.17*	–0.16*	0.11	0.04	0.18**
Pressure group activity	0.12	0.08	0.15*	0.03	–0.01	0.01	0.05	0.18*	–0.05
Change of majority	–0.02	0.02	–	–0.15	–0.13*	–	–0.14	0.02	–
Organizational									
INFO systems	0.37**	–0.04	–0.01	0.34*	0.02	–0.04	–0.04	–0.07	0.001
Personal									
Age (mayor)	–	–	–	0.04	0.06	0.05	–0.08	–0.03	–0.06
Education (mayor)	–	–	–	0.03	0.14*	0.01	–0.21	0.09	0.11*
Contextual									
Fiscal slack	–0.02	–0.10	–0.09	0.06	0.001	0.10	0.02	0.04	0.02
Size	–0.04	–0.06	0.05	0.11	–0.02	0.11	0.05	–0.09	0.03
Min. *N*	95	242	266	37	235	213	95	241	266

* Significant at the 0.05 level.
** Significant at the 0.01 level.

Leadership styles   A paternalistic leadership style is associated with parties of the left, not parties of the right, as initially expected. Only the USA demonstrates the relationship, and not very strongly. It is surprising to find it there, though, where the two major parties are often thought of as more similar than the major protagonists in European countries. In Norway it is pressure groups that seem to produce paternalism among mayors. Personal variables are not associated with leadership style, with the possible exception of education in the US case. Size does not have the expected effect: it is unrelated to paternalism or populism.

CONCLUSION

Levels of fiscal stress or slack do not affect city politics and leadership patterns very much – unless levels of fiscal stress are very high, as in some US cities. That is the one conclusion that stands out most clearly following the data analysis presented here. There are other city-level variables that seem more important, such as political party affiliation, pressure group activity, organizational sophistication or even personal qualities, to some extent. However, these variables are not uniformly important or of the same effect across municipalities in the three countries studied here. It seems that system characteristics shape local politics and leadership more than city-level variations within countries. But many of these system-level variations are different from what was expected on the basis of the literature of previous research.

We therefore need to explore and compare the characteristics of our local government systems more closely than we have done so far if we are to be able to understand and predict the course of local politics and leadership better than at present.

NOTES

I wish to thank Sissel Hovik and Ishtiaq Jamil for their assistance in data preparation for this chapter.

1 Time of surveys: France, 1985; Norway, 1985/86; USA, 1983/84. Response rates for managers: France, 46 percent of 381 units; Norway, 82 percent of 454 units; USA, 49 percent of 1,030 units. Response rates for mayors are about the same. My thanks to American and French teams for permission to use their data.

REFERENCES

Baldersheim, H. 1982. 'Administrative Leadership in a Big City', in G.M. Hellstern et al. (eds), *Applied Urban Research: Proceedings of the European Meeting on Applied Urban Research*, vol. II, Essen: Bundesforschungsanstalt für Landeskunde und Raumordnung.

Baldersheim, H. 1986. *Men har dei noko val? Styrings- og leiingsprosessar i storbykommunane*, Oslo: Universitetsforlaget.

Baldersheim, Harald and Torodd Strand 1988. '"Byregjering" i Oslo: Hovedrapport fra et evalueringsprosjekt', Rapport 88/7, Bergen: LOS-senteret.

Baldersheim, H. et al. 1989. 'Free to choose? The Case of Affluent Norwegian Municipalities', in S.E. Clarke (ed.), *Urban Innovation and Autonomy: The Political Implications of Policy Change*, Sage Urban Innovation Series, vol. 1, London: Sage.

Brofoss, K.E., A.T. Cowart and T. Hansen 1975. 'Budgetary Strategies and Success at Multiple Decisions Levels in the Norwegian Urban Setting', *American Political Science Review*, 69: 231–247.

Brunsson, Nils 1983. *Går det att spara?* Stockholm: Doxa förlag.

Brunsson, Nils 1985. *The Irrational Organization*, Chichester: John Wiley.

Clark, T.N. 1985. 'The Dynamics of Political Culture: Liberalism, Radicalism, and the New Fiscal Populism', *Toqueville Review*, 7: 174–190.

Clark, T.N. 1987. 'Defending City Welfare: How are the Hard Choices Made?' in *Key-Note Lectures from an International FAUI Seminar at LOS-senteret*, Bergen: LOS-senter Notat 88/10.

Clark, T.N. and L.C. Ferguson 1983. *City Money*, New York: Columbia University Press.

Clark, T. N. et al. 1987. 'Political Culture and Urban Innovation', paper presented at the annual meeting at the American Sociological Association, Chicago.

Clarke, S.E. (ed.) 1989. *Urban Innovation and Autonomy: The Political Implications of Policy Change*, Sage Urban Innovation Series, vol. 1, London: Sage.

Czarniawska-Jorges, Barbara 1988. 'Organizational Change: Possible Perspectives', Notatserie, LOS-senteret, 13/88.

Dente, B. and F. Kjellberg 1988. *The Dynamics of Institutional Change: Local Government Reorganization in Western Democracies*, London: Sage.

Eulau, Heinz 1971. *Policy-Making in American Cities: Comparisons in a Quasi-Longitudinal, Quasi-Experimental Design*, New York: General Learning Press.

Goldsmith, Michael 1987. 'The Changing Role of Mayors and Chief Executives', paper presented at the ECPR workshop on Leadership and Local Politics, Amsterdam.

Kotter J., and P. Lawrence 1974. *Mayors in Action*, New York: Allen and Unwin.

Larsen, Helge. O. 1990. 'Ordføreren – handlekraft eller samlende symbol?' in H. Baldersheim (ed.), *Ledelse og innovasjon i kommunene*, Oslo: Tano forlag.

March, James G. 1984. 'How We Talk and How We Act: Administrative Theory and Administrative Life', in T.J. Sergiovanni and John E. Corbally (eds), *Leadership and Organizational Cultures*, Urbana: University of Illinois Press.

Midwinter, Arthur 1988. 'Local Budgetary Strategies in a Decade of Retrenchment', *Public Money and Management*, autumn: 21–28.

Morgan, R.D. and W.J. Pammer, Jr 1986. 'Responding to Urban Fiscal Stress: the Case of Arkansas and Oklahoma', in T.N. Clark (ed.), *Research in Urban Policy*, vol. 2, pt B.

Mouritzen, P.E. and K. Houlberg Nielsen 1988. *Handbook of Comparative Urban Fiscal Data*. Odense: Danish Data Archives.

Page, Edward and Michael Goldsmith 1987. *Central and Local Government Relations*. London: Sage.

Peterson, Paul E. 1981. *City Limits*, University of Chicago, Press.

7

Choosing the Budget Size: Mayors' Spending Preferences in a Cross-National Perspective

Richard Balme

In order to assess the implications of the fiscal situation (stress or slack) for political leadership, we need to consider leaders' preferences for different policies, to understand under which circumstances mayors are more or less pro-spending-oriented, and in which government areas these preferences are more intense.

Leaders may have different views about the budget's ideal size, for instance according to ideological beliefs for big or small government, for redistribution or market-oriented policies. Mayors also have to establish priorities in dealing with the pressures of day-to-day government and defining solutions to conflicts of interests and urgent situations. Leaders' preferences concerning cities' budgets may also vary across cultures and are shaped through the policy process.

This chapter is an attempt to map these preferences and to understand their origins. We address four specific questions:

1 Do leaders' preferences differ across areas of city government?
2 Do patterns of preferences differ across countries?
3 Are mayors' item-by-item spending preferences consistent with their general fiscal policy orientation, liberal or conservative, anti- or pro-spending?
4 What are the factors – ideological, institutional and political – shaping leaders' preferences?

To the first two questions the answer is yes; in particular, cross-country differences are more important than expected. To the third the answer is more ambiguous, as consistency of preferences relies heavily on the assumptions retained in their evaluations. Finally, a first assessment of origins of spending preferences reveals no single determinism, but suggests different equilibria between multiple variables defining national patterns.

Mayors were asked to report their spending preferences (more, less, or the same level of spending) on a set of issues in each country. Data are

available for the USA, France, Denmark, Norway and Finland.[1] Items
sometimes differ from country to country, partly because of variations
in functional responsibilities, partly because of constraints on the survey
in each country.[2] Nevertheless, questionnaires provide a consistent survey
item concerning leaders' spending preferences collected cross-nationally.
Our purpose is first to describe patterns of preferences by government
functions and by country. We then comment on the consistency of
preferences revealed by the data. Finally, we try to explain these patterns
of preferences and to identify their determinants.

PUBLIC SPENDING TO BUY WHAT?

Budget items are not necessarily relevant for political theory. If properties
of policy outputs shape the choice of decision-makers, then budgetary
categories have to be classified according to different policy types.
We know from Samuelson's classical analysis (Samuelson, 1954) that
public spending can buy two different categories of goods: separable
goods called private goods, and non-separable or public goods, where
jointness of supply and non-exclusion of consumption make optimal
provision problematic. Neoclassical economics sometimes assumes that
private goods will be efficiently provided through market mechanisms,
while the state should supply pure public goods such as lighthouses,
clean air and national defense to overcome free-riding and collective
action problems. In fact, governments always buy both categories of
goods. Public opinion studies have reported important differences in
the determinants of citizens' preferences for public and private goods
in city governments (Hoffman and Clark, 1979). The order of priorities
in leaders' spending preferences may vary along the same classification as
their constituencies differ, and because these types of policy outputs have
different political implications.

At first sight, pure public goods appear to be the 'core' of the minimal
state, and leaders might prefer to spend more on public than on private
goods. On the contrary: in A. Hirschman's terminology (Hirschman,
1970), 'loyalties' initiated through public spending often involve the
patronage and clientelism associated with private goods. Without exclusion mechanisms, pure public goods benefit everybody, voters and
non-voters, supporters and opponents. Such policy outputs are unable
to reward political support, as free-riders enjoy their advantages along
with everyone else. In the competition for political positions, the search
for political support is likely to orient leaders toward public provision of
private goods, generating alternative policy coalitions. Such a tendency is
viewed by Wildavsky (1980) as a major cause of government growth, and
by Olson (1982) as generating economic decline after long-term periods
of prosperity. Along these lines, the welfare state crisis is interpreted as

a consequence of the pluralist structure of interests and the competition for power. Given the importance of welfare programs in public budgets, decision-makers' spending preferences may be higher for allocation or redistribution policies than for more traditional functions of government.

If the difference in the intensity of preferences is difficult to anticipate, it seems reasonable to hypothesize that volatility will be higher concerning private goods, owing to variation in ideological beliefs and in the necessity to mobilize support, and because of the adaptive capacity of supply. Public goods spending preferences are expected to be more constrained, because they generally involve more basic functions of government, in the sense that there is no alternative mechanism of provision, and because lumpiness makes progressive supply adjustment difficult.

To test these propositions, we classified cities' spending items in three broad categories: allocation policies, urban development policies and municipal management. The most common classification of government functions is between redistribution, allocation and development (see for instance Lowi, 1969 or Peterson, 1981). We did not retain this classification because redistribution and allocation policies are extremely difficult to separate in terms of spending, and because there is an overlap between municipal management and other categories; for instance, evolution in the number of municipal employees can be considered both as a development of urban services and as an allocation policy. Instead, we defined a general category for allocation policies including both allocation and redistribution. The principle defining this category is that the service should be a private good, in the sense that citizens can choose to avoid its consumption, and that an exclusion mechanism can be created to select the beneficiaries. This category includes education, hospitals, welfare services and low-income housing, among others. Urban development policies include more public goods, in the sense that benefits are (at least in principle) widespread among the entire population: such services do not display exclusion mechanisms. They include streets and parking, mass transportation, fire protection, public utilities and economic development.

The third category concerns municipal management. Preferences about internal organization and management include capital stock and investment, administration and planning, the number of municipal employees and their salaries. As suggested above, such items correspond both to private goods for bureaucrats or employees and to public goods (or costs) for citizens, if we admit that government benefits everybody. Of course, government activity benefits some interests and hurts others, and rarely affects citizens equally. But at least some of these benefits are quite universally shared, such as fire protection. Municipal management expenditures involve paying for the private goods necessary to public goods provision. Mayors are expected to be more fiscally liberal on municipal management, because of the primacy of such spending on other policies, and because of responsiveness to municipal employee pressure.

For each of these items and categories, a mean score was calculated for each country. These scores were then converted to a *percentage difference index* ranking from −100 to +100. Using this method gives scores a substantial meaning. Thus, a score of +25 means that the proportion of respondents preferring more spending minus the proportion preferring less spending is 25. Details for category building are specified in the footnotes to each table.

Do mayors want more or less spending? To a large extent, it depends on how we ask them (see Table 7.1). If we compute the mean of their scores for each item, they want to spend more in the USA, Finland, France and Norway, and slightly less in Denmark (row 1). However, these preferences are not necessarily identical with their choice when asked how much they want to spend in all areas of city government (row 2): American and Finnish mayors are consistent, but French officials display opposite choices: they want a little more of everything, but would like overall spending to be reduced.

Previous studies have shown that such a contradiction may be produced by the scope of choice proposed by the survey. If the domain of preferences is unrestricted, respondents may answer without limitation on an item-by-item basis. This myopia would forbid the summing-up of these preferences without restricting the scope for choice, for instance through tools such as budget pies (Clark, 1974; McIver and Ostrom, 1976). Should we then take this result as a bias of the survey technique? Probably not, as two survey results out of three do not display such a contradiction. Mayors are provided with more information than anyone else and should not be abused by the questionnaire. They are the most decisive executives in the budgetary process, and it seems reasonable to assume that they know what they want.

Does this reveal an inconsistency of mayors' preferences? The point is more theoretical. When asked about all areas of city government, French mayors would express wishful thinking. Faced with issue-by-issue choices, they would then be responsive to the emergency of situations, and would give way to incrementalist decision-making. An alternative explanation is that mayors weight differently the items in the questionnaire, which makes sense if we consider that the budget of parks and recreation has a very different size from the one of personnel expenditures. Thus, French mayors want to increase spending on most items, but would like large cutbacks in a few areas (personnel expenditure reduction) to allow for a general decrease in total expenditures. Such an expectation is questionable, but not unrealistic if we consider that personnel expenditure accounts for 43 percent of French cities' current

Table 7.1   *Mayors' spending preferences (mean scores)*[1]

	USA	Fr	Dk	No	Fi
Mean, all items[2]	20	13	–5	21	30
Preferences, all areas[3]	12	–23	NA	NA	30
Allocation policies					
Education	36	16	–12	5	19
Public health/hospitals	5	20		29	31
Low-income housing	11	43		18	47
Social Welfare[4]	–18	37	–7	16	33
Sports/leisure/culture		13	13	28	30
Libraries			–5		
Mean score	8	25	–3	19	32
Urban development policies					
Streets and parking	47	40	29	44	32
Mass transportation	17	3	–22	–23	
Parks and recreation	15			9	26
Police/public order	39	23			16
Fire protection	14			–1	17
Garbage collection		–2			
Public utilities			–2	28	
Economic development		69		45	43
Environment protection					33
Housing					31
Mean score	26	26	1	17	28
Municipal management					
Capital stock	59	21			
Investments					35
Administration/planning			–23	24	27
No. of employees[5]	–8	–39		9	28
Wages of employees[5]	33	–39		0	28
Mean score	28	–9		21	30

[1] Scores are means converted to a percentage difference index ranging from –100 to +100.

[2] Means displayed here are calculated across all items for each country.

[3] In some countries, a specific question was asked about general spending preferences for all areas of city government.

[4] Includes for Denmark day-care institutions, elderly programs, unemployment programs, social cash benefits, social counselling. For Finland: children and youth, elderly and other social services. For other countries, only the general item 'welfare spending' was retained.

[5] Some countries specified wages and number of municipal employees. For other countries, the score for the general item 'municipal employees' is reported twice in each of these two categories.

spending.[3] In any case, such a preference pattern displays a conflicting aspiration for the provision of more efficient government outputs.

What then do we know from these answers? First of all, we should be cautious in interpreting the results as the calculus of a mean with unweighted items is a rather strong assumption, as budget shares are not equal for each item. Inconsistencies between attitudes item for item and attitudes toward total expenditures have already been addressed (Wildavsky, 1980; Kristensen, 1980). We agree that more detailed studies should weight preferences according to the proportion of spending on each item in the budget (Kristensen, 1982). Such an operationalization is nevertheless difficult for our purpose. Functional responsibilities may vary quite a lot in some countries. Therefore we would need to know budget structures at the city level, and these are unfortunately unavailable at this point. Our findings suggest, first, that such inconsistencies are not systematic; second, that American and Finnish mayors definitely want a larger budget, and that French mayors want a different distribution of spending; and, third, that Danes want less spending on most items and Norwegians, generally more. Note that in every case, these tendencies are rather small, and that no country displays a strong anti- or pro-spending orientation. The general picture is that mayors favor a small increase in cities' budgets except in Denmark, where they wish to stop budgetary growth.

How are these preferences distributed among government functions? According to Figure 7.1, there is no consistent pattern. Development policies are neither necessarily preferred, nor necessarily sacrificed to allocation policies. Contrary to our expectations, preferences for municipal management policies are not more heavily constrained than for other items. Among all categories, management presents the larger deviation, from −23 in Denmark to +30 in Finland. Mayors do not appear to be hostages of the municipal bureaucracy.

True, spending preferences for public goods vary less than for private goods or municipal management, and mayors do not want to reduce such spending. The public-versus-private goods categories are then relevant to explain the scope rather than the orientation of policy choice. In other words, there is more consensus among mayors across nations concerning development policies than allocation policies. But there is no regular pattern systematically favoring public over private goods provision or the reverse. We find some support for the hypothesis concerning the difference in volatility of spending preferences; but we reject the hypothesis concerning the difference in intensity of spending preferences between private and public goods. Municipal management does not appear as a policy domain more heavily constrained than others, and does not necessarily rank higher or lower than other categories of expenditures.

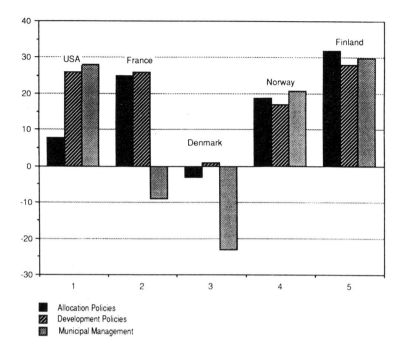

Figure 7.1   *Patterns of mayors' spending preferences: intensity of mayors' spending preferences as a percentage difference index*

CROSS-NATIONAL PREFERENCE PATTERNS

This cross-national examination shows that policy types as defined here are not sufficient to predict mayors' spending preferences. Other factors must be considered. In particular, political cultures might define policy preferences: variations in partisan ideologies, social cleavages, cultural traditions and institutional arrangements are likely to produce different patterns of preferences in different countries. Our data suggest that such differences are important. This section describes in greater detail patterns of preferences for each country.

American mayors are selective in their pro-spending orientation. They want more municipal management and more urban policies for the same level of allocation. Remarkably, they favor more spending on education and less on social welfare, which might reflect a preference for equality of opportunity rather than achievement, and probably the saliency of the public school issue in the USA, as we are here dealing with decision-makers' preferences. Note that European mayors favor more spending on all allocation items, except for the Danes, who prefer less on most of them (the exception being sports, leisure and culture).

The US model is interesting as it displays preferences for more government with a public goods orientation. Norwegians and Finns present a classical social democratic orientation: they want more government in every category, with the priority given to allocation policies. By contrast, mayors from Denmark want nearly the same level of government for a better price, by reducing administration spending. Following the same orientation toward productivity, French mayors generally want more government for less administration; they want to increase allocation and development expenditures while reducing personnel expenditures. Again, French mayors seem to display the most conflicting pattern of preferences.

If we look at priorities among these preferences, the picture gets more precise: US mayors mainly want to implement cutbacks on welfare spending, and especially wish to develop capital stock. The French and Norwegians give priority to economic development, whereas the Finns are concerned with low-income housing and the Danes with streets and parking. As opposed to other countries, Danish and French mayors display a strong productivity orientation, as they wish to reduce administration and personnel expenditures. It seems that the US conception of local government has no real equivalent in Europe among the countries selected. The French, Finns and Norwegians favor the development of the local welfare state with a productivity orientation for French mayors. Danish mayors want to stop the development of welfare spending and improve the productivity of administration. In Norway and Finland, the classical social democracy is not questioned by mayors. In other European countries, the ratio between government inputs and outputs is perceived as conflicting. But US mayors are the only ones wanting to provide more government without increasing allocation expenditures.

EXPLAINING POLICY PREFERENCES

Where do cross-national differences in spending preferences come from? Following Mouritzen (1987), policy preferences are a function of three main variables: policy outputs, self-interest, and ideology. Mayors' spending preferences are expected to depend on previous policy outputs, reflected through different levels of fiscal stress and tax efforts. 'Self-interested' mayors are basically concerned with elections: their preferences are defined primarily by their responsiveness to pressure groups and to what they perceive as their constituents' preferences. Organized groups' activity, including municipal employees pressure, might shape decision-makers' spending preferences. If the electoral process is efficient, citizens' preferences should also be taken into account and at least partially reflected by leaders' policy orientations.

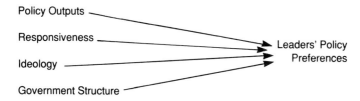

Figure 7.2   *A model of the formation of leaders' preferences*

Alternatively, the formation of leaders' preferences may be more ideological, for instance according to the partisan affiliation of mayors. Ideology would then override responsiveness in policy choice, and left–right cleavages would determine policy preferences, if not outcomes. Finally, government structure is likely to affect spending preferences: mayors are expected to be more pro-spending-oriented in centralized than in decentralized countries, as they can rely on central grants to support their policies. These relations are depicted in Figure 7.2.

The hypotheses suggested above are considered through a set of first tests, in the hope of refining their formulation for further research. We first assess a seemingly very basic cause: the budget itself. It has already been suggested that mayors may be more fiscally conservative when fiscal

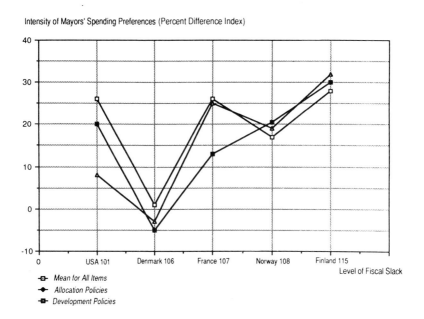

Figure 7.3   *Mayors' spending preferences at different levels of fiscal slack (Mouritzen and Nielsen, 1988: 39)*

Mayors' Spending Preferences (Percent Difference Index)

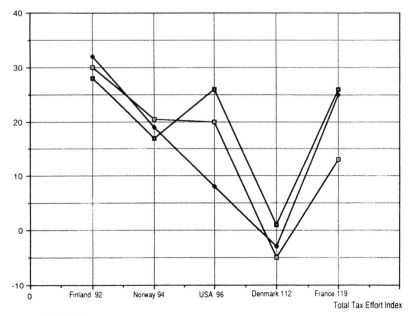

-■- *Mean for All Items*
-●- *Allocation Policies*
-■- *Development Policies*

Figure 7.4   *Mayors' spending preferences at different levels of tax effort (Mouritzen and Nielsen, 1988: 30)*

conditions are more difficult and city governments more autonomous. How do the data fit with these propositions?

Figure 7.3 indicates the association between fiscal slack in the previous period and mayors' spending preferences.[4] There seems to be a slight tendency for mayors to prefer more spending in countries where slack is higher. Nevertheless, Danish mayors, with a fiscal situation close to that of France or Norway, clearly prefer less spending, and this suggests the need for other explanations. Among the different policies, allocation spending (private goods provision) seems to follow the most linear relation with fiscal slack. Tax effort, reported in Figure 7.4 exhibits a similar trend.[5] Note that France, an outlier in the distribution, is also the country where mayors prefer the largest reduction in personnel expenditures, apparently without equivalents in other countries.

Contrary to our expectations, centralization, measured through the percentage of total grants in cities' budgets, does not affect leaders' spending preferences at the aggregate level.[6] Finally, organized groups' activity considered at the country level seems to be a reasonable predictor

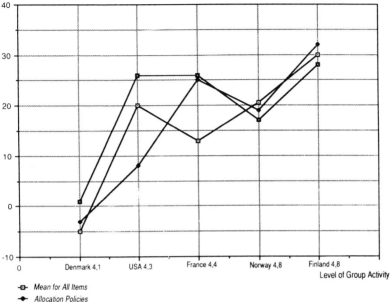

Mayors' Spending Preferences (Percent Difference Index)

Level of Group Activity

-□- Mean for All Items
-◆- Allocation Policies
-■- Development Policies

Figure 7.5 *Mayors' spending preferences at different levels of groups' activity*

of mayors' spending preferences. Figure 7.5 exhibits an apparent linear relation between both variables. Note that preferences for public and private goods do not seem to follow the same curvature. Preferences for spending on private goods are more continuous and monotonous, whereas those for spending on public goods seem to follow a decreasing marginal rate of growth.

Naturally, we have too few points of observation to attribute any statistical significance to these relations. The findings mainly suggest new research directions. They bring some support to the hypothesis of the effect of policy outputs and organized groups' activity on leaders' preferences, without validating at the country level the hypothesis relating to government structure. Some examination at the city level gives an insight to these developments.

Indeed, analyses performed at the city level do not reveal that the same relations are equally active everywhere. Table 7.2 displays results from tests using bivariate analysis in the USA, France and Norway. In none of these countries was fiscal stress found to be a significant

Table 7.2   *Association between mayors' spending preferences and major indicators at the city level*

	USA	France	Norway
Fiscal stress			
Tax effort			−
Fiscal dependence		+	+
Critical environment[1]	+		+
Group activity[2]	+		+
Citizens' preferences[3]	+	+	+
Leftist government[4]		+	+

Empty cells indicate no significant effects. + and − indicate direction of correlation significant at 0.01.

[1] Intensity of problems in the city as reported by mayors.

[2] Intensity of organized groups' activity as reported by mayors.

[3] Citizens spending preferences as perceived by mayors.

[4] For the USA, percentage of Democrats among council members and partisan affiliation of mayors; for France, partisan affiliation of mayors; for Norway, partisan affiliation of mayors and votes for leftist parties.

Source: Results from this table concerning the USA are reported from Balme and Miranda (1988); they have originally been computed for France and Norway for the purpose of this book.

influence on mayors' spending preferences. Only in Norway does tax effort limit mayors' preferences. Fiscal dependence, measured through the percentage of grants in the cities' budgets, exhibits the expected effects in France and Norway: mayors are more pro-spending-oriented when grants are more important. Grant allocation criteria, which are generally based on a needs evaluation, are probably reflected in the mayors' spending preferences. Group activity and a critical perception of their environment orient mayors in favor of spending in Norway and in the USA, but not significantly in France.

Finally, different indicators of leftist political culture (mayor's partisan affiliation, votes for pro-spending party, leftist representation in municipal council) suggest that Norwegian and French mayors are more spending-oriented when they belong to leftist parties, but such a difference is not significant in the USA. This confirms and expands an earlier result by Miranda and Clark (1988). This set of findings seems to describe US mayors' preferences as being driven by local demand through pressure groups and a perception of their environment. On the other hand, French mayors seem more influenced by exogenous factors, such as partisan affiliation and fiscal dependence. Finally, Norwegian mayors exhibit an accomplished model of local social democracy, as their preferences are both responsive to pressure and shaped by political parties. Policy preferences are more locally defined in the USA and more partisan in Europe. But French mayors are apparently more strictly ideological than their Norwegian counterparts.

POLICY MAPS: CITIZENS' PREFERENCES EVALUATED BY MAYORS

To what extent do leaders feel constrained by their constituents in their budgetary decisions? Mayors were asked to evaluate citizens' preferences on a given set of issues. We first describe these patterns of preferences as they appear reported by mayors (Table 7.3). The most apparent result is that mayors do not perceive citizens as opposed to budgetary growth. The same remarks about consistency of aggregate preferences measured through unweighted indicators hold here, but using means for all items reveals an orientation for more spending. A direct survey of citizens in Denmark leads to the same finding.[7] Citizens may feel more or less pressured by taxes, but in general they still want more services. The point is not new, and regularly appears in surveys and public opinion studies (for instance Citrin, 1979; Smith, 1987b). Changes in spending preferences, unfortunately unavailable in our study, should be considered if we want to evaluate public opinion impact. Nevertheless, what matters here is that, whatever the evolution, mayors still perceive people as wanting more government rather than less.

A 'world tax-revolt' would need to be supported by a systematic tendency of constituents to support a reduction in the cost of government, and therefore to prefer less spending for municipal management. As opposed to this view, Finnish mayors think that voters want to increase expenditures on such items, including municipal employment. This result shows that a growth in the size of the budget is not necessarily a 'public bad', and that, beyond services, benefits are also expected from government's activity itself. Along the same line, are citizens orientated more toward private benefits than public goods? In other words, do mayors feel more demand for spending on private than on public goods? Not systematically: according to the data, such a relation is true for European countries, but not for the USA, where mayors perceive citizens as favoring development policies, but wishing to reduce expenditure on allocation policies (Figure 7.6). Characteristics of policy outputs do not strictly determine spending preferences, and different countries exhibit different patterns.

These patterns generally correspond to mayors' own preferences, with a few specificities: Finland is the only country where citizens are reported to favor municipal management, basically through the large number of municipal employees. In other countries, local bureaucracy and municipal employees are perceived as expensive, and mayors feel pressured by voters to reduce their numbers. Assuming that mayors are right in their

Table 7.3　*Estimations of citizens' spending preferences by mayors (mean scores)*[1]

	USA	Fr	No	Fi
Mean, all items[2]	7	23	12	36
Preferences, all area[3]	−33	−45	37	
Allocation policies				
Education	0	51	13	18
Public health/hospitals	−16	41	38	35
Low-income housing	−43	51	20	49
Social welfare[4]	−51	58	35	42
Sports/leisure/culture		33	36	38
Libraries				
Mean score	−28	46	26	36
Urban development policies				
Streets and parking	30	61	43	42
Mass transportation	−9	12	−10	
Parks and recreation	7		−1	31
Police/public order	53	54		23
Fire protection	53		−2	21
Garbage collection		8		
Public utilities			24	
Economic development		66	41	36
Environment protection				40
Housing				44
Mean score	21	40	16	34
Municipal management				
Capital stock	20	17		
Investments				38
Administration/planning			−7	47
No. of employees[5]	−53	−78	−40	52
Wages of employees[5]	38	−78	−29	52
Mean score	−24	−31	−25	47

[1] Scores are means converted to a percentage difference index ranging from −100 to +100.

[2] Means displayed here are calculated across all items for each country.

[3] In some countries, a specific question was asked about spending preferences for all areas of city government.

[4] For Finland, includes children and youth, elderly and other social services; for other countries, only the general item 'welfare spending' was retained.

[5] Some countries specified wages and number of municipal employees; for other countries, the score for the general item 'municipal employees' is reported twice in each of these two categories.

Intensity of Spending Preferences (Percent Difference Index)

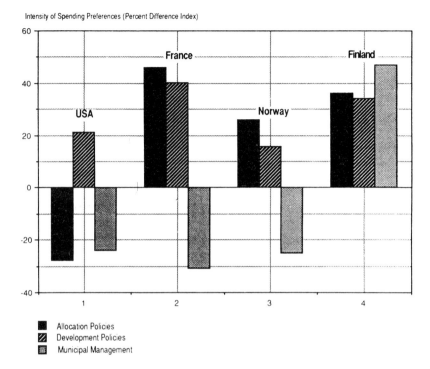

Allocation Policies
Development Policies
Municipal Management

Figure 7.6 *Citizens' spending preferences as evaluated by mayors*

evaluation, the Norwegians and the French want more spending both on welfare and development.

Americans, finally, display the most specific pattern, with preferences for more development policies and less allocation. Among these, mayors perceive citizens to favor cutbacks in welfare programs but not in education. This specificity holds both for mayors' own preferences and for their evaluation of citizens' opinions. It provides for local governments a validation of cultural differences in support for the welfare state, already pointed out at the national level (Coughlin, 1980; Smith, 1987a). Such a tendency has been interpreted by Wilensky (1975) as an orientation of European political cultures toward a greater equality of achievement. US values would not support such welfare spending, but nevertheless would be concerned with education spending, corresponding to equality of opportunities. This interpretation fits the general idea of US individualism and exceptionalism.

However, US mayors are probably excessive in their evaluation when they perceive citizens as preferring to reduce welfare spending, as this perception is not directly consistent with the polls. Gallup, for instance, never found a majority of citizens in favor of reduced welfare spending.[8] The point is important: if mayors were right in their evaluation of citizens' preferences, voters would display opposite patterns of choice for local and federal spending. Such a discrepancy would not be inconsistent in the USA where city governments have a comparatively strong autonomy. However, this explanation needs to be confirmed through direct citizens' surveys in the USA. If citizens' preferences happened to be consistent at the local and federal level, then a 'fiscal illusion' would not necessarily be on the side of voters. In other words, leaders' evaluation of public opinion may be biased, either as a result of media pressure or as a projection of their own preferences to reduce uncertainty about their constituents' tastes.[9] Baldassare (1986), testing this proposition with data collected in Orange County, California, reported an underestimation of citizens' spending preferences for education and public health in mayors' perceptions.[10] What matters here is not the actual preferences of citizens, but their perception by the mayors. Whether they are right or wrong, US mayors *feel* driven by their constituents to reduce welfare spending, and appear clearly distinctive from their European counterparts on this point.

Differences between mayors' own preferences and the ones they attribute to citizens define policy distance. We depict the policy distance in Table 7.4 as the difference between mayor's preferences (Table 7.1) and their evaluation of citizens' preferences (Table 7.3). A positive score indicates that mayors feel more pro-spending than citizens.

Mayors feel more or less responsive to their constituents in different countries and about different issues. In every country where such data are available, personnel spending is among the items where the discrepancy is the largest, although in Finland leaders feel pressured by their constituents to hire more employees, contrary to other countries. The highest discrepancy in relation to wages of municipal employees occurs in the USA, where leaders differ from citizens, the latter wishing to lower the costs of service provision. Note also that European mayors want to spend less than their constituents on allocation policies, while US mayors want to spend more. Analysing policy distance suggests different patterns of policy agendas (see Figure 7.7). With support for local expenditures, the discrepancy between leaders and citizens may vary. The dynamic element of these policy agendas may be leaders or, alternatively, citizens. Can we talk about a 'spending revolt' in the same way that we view a tax revolt? Obviously not, as citizens' preferences in every country seem to favor growth rather than limitation of local government expenditures. Is this revolt a leaders' revolt, opposing budgetary incrementalism driven by citizens' preferences? This seems to be the case in Denmark but not in other countries, where mayors favor expenditure growth. The revolt, if any, occurs among leaders, and is limited to Denmark.

We learn from these data that public opinion, as judged by the political

Table 7.4 *Policy maps: Difference between mayors' and citizens' spending preferences*[1]

	USA	Fr	Dk	No	Fi
Mean, all items[2]	13	−10	−12	9	−7
Allocation policies					
Education	36	−35	−18	−8	2
Public health/hospitals	21	−21		−9	−4
Low-income housing	54	−8		−2	−2
Social welfare[3]	33	−21	−25	−19	−10
Sports/leisure/culture		−20	10	2	−8
Libraries			−4		
Mean score	36	−21	−9	−7	−4
Urban development policies					
Streets and parking	17	−21		2	−10
Mass transportation	26	−9	−38	−13	
Parks and recreation	8			10	−5
Police/public order	−14	−31			−8
Fire protection	−39			1	−4
Garbage collection		−10			
Public utilities			45		
Economic development		3		5	8
Environment protection					−7
Housing					−12
Mean score	−2	−14		1	−6
Municipal management					
Capital stock	39	4			
Investments					−2
Administration/planning				30	−19
No. of employees[4]	45	39		48	−24
Wages of employees[4]	71	39		29	−24
Mean score	52	27		36	−17

[1] Scores are computed as differences between scores from Tables 7.1 and 7.3. For Denmark, citizens' spending preferences are from a direct survey of citizens reduced to a small number of items. In the percentage difference index, scores from this survey are: Education +6; Social welfare +18; Sports/leisure/culture +3; Libraries −1; Mass transportation +16; Rebuilding inner city +1.

[2] Means displayed here are calculated across all items for each country.

[3] For Denmark, includes day-care institutions, elderly programs, unemployment, social cash benefits, social counselling; for Finland: children and youth, elderly and other social services; for other countries, only the general item 'welfare spending' was retained.

[4] Some countries specified wages and number of municipal employees; for other countries, the score for the general item 'municipal employees' is reported twice in each of these two categories.

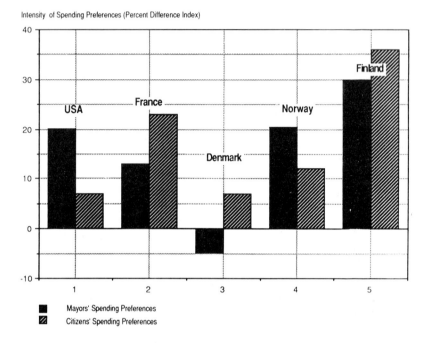

Figure 7.7   *Patterns of policy distance in five countries*

leaders, still favors local government growth. Does this mean that public spending will keep on growing everywhere under the demand for a welfare state? Apparently not, as in some countries, like the USA or Norway, mayors feel more spending-oriented than citizens; the pressure of citizens' preferences for public spending might be declining, and could progressively induce leaders to freeze the level of expenditures. In France and Finland, however, a classical pattern of welfare state support seems to be active, with leaders following and, in a way, moderating citizens' demand for policy outputs.

CONCLUSION

We briefly summarize the patterns of spending preferences found in this chapter. In the USA, leaders feel that they are more fiscally liberal than their constituents and favor a 'public good' expansion of local government. This can be contrasted with European models, where welfare spending at the city level is questioned. French mayors feel pressured by their constituents to increase development and welfare spending at a lower cost. Danish mayors are more fiscally conservative than their constituents:

they revolt against the expansion of local government and wish to lower its costs. Norwegian and Finnish politics are more traditionally social democratic. Norwegian mayors feel more spending-oriented than their citizens, especially on personnel expenditures. Finally, Finland presents a rather specific case where mayors feel pressured by citizens to develop the welfare state, including more employment by city government. The US model of leaders' preferences is more distinctive than expected – less ideological than the European model, responsive to local rather than national political factors, and to what is perceived as an opposition of citizens to welfare spending at the city level.

These patterns mapped out, we do not argue that preference formation is totally idiosyncratic and embedded in national 'temperaments'. We do believe that policy preferences depend on policy output characteristics and ideological factors, the effects of which were shown above. However, the combination of these elements remains complex and gives rise to distinctive patterns in different countries. The contrasts displayed here should help us to consider some of these relations. Further research should first collect new data. We need to compare local and nationwide citizens' preferences, and to develop a better understanding of leaders' assessment of citizens' preferences. Probably the most important need is to compare these preferences over time. More feasible in the near future, the development of causal analysis of spending preferences at the city level will probably bring new results concerning socio-economic determinants of preferences and their effects on political leadership and, finally, on policy outputs.

NOTES

1 The Danish data cover the political elite comprising mayors, members of the influential finance committee and chairmen of the standing committees.

2 About this methodological issue applied to the field, see Smith (1987b) and Rasinski (1989).

3 Percentage of permanent personnel spending in cities over 10,000 in 1985 (Ministère de l'Intérieur, 1987).

4 Fiscal slack is measured for the period immediately before the survey. For Denmark, where the survey was performed in late 1981, we retained fiscal slack 1979–82; for other countries, where data were collected in 1984 or 1985, we used fiscal slack 1982–84.

5 Total tax effort index in 1984 with 1978 as base year, except for Denmark, where tax effort is 1981 with 1978 as year base. Our tax effort index thus measures change in tax effort over the period leading up to the survey of mayors' spending preferences. Tax effort is here defined as total municipal taxes divided by the tax base.

6 We do not display the relevant graph for reasons of space.

7 Inconsistencies in citizens' preferences may differ according to ideology: conservative voters would be most inconsistent, wanting lower taxes without

reduced spending. Leftist voters would be more pro-spending, but not as critical as conservative voters about the level of taxation. This remark is due to P.E. Mouritzen, who referred to a citizens' survey in Denmark.

8   For instance, Gallup Report no. 234 (March 1985) only found 22 percent of Americans thinking that there was too much social spending, and only 41 percent supporting cutbacks in welfare programs to reduce the deficit. The report shows that defence cutbacks would be approved in such a case by 61 percent, and that these results were consistent with earlier findings.

9   About biases in such estimations, see Fields and Schuman (1976) and Dekker and Ester (1989).

10   Unfortunately, this survey does not provide a specific category for welfare spending.

REFERENCES

Baldassare, Mark 1986. 'How Well Do Mayors Assess Citizens' Preferences?' *Research in Urban Policy*, 2: 159–174.

Balme, Richard and Rowan Miranda 1988. 'Political Leadership and Urban Innovation', paper presented at the Midwest Political Science Association meeting, Chicago, April.

Citrin, Jack 1979. 'Do People Want Something for Nothing? Public Opinion on Taxes and Government Spending', *National Tax Journal*, 32(2): 113–129.

Clark, Terry N. 1974. 'Can You Cut a Budget Pie?' *Policy and Politics*, 3: 3–31.

Coughlin, Richard M. 1980. *Ideology, Public Opinion and Welfare Policy*, Berkeley: University of California Press.

Dekker, Paul and Peter Ester 1989. 'Elite perceptions of Mass Preferences in the Netherlands: Biases in Cognitive Responsiveness', *European Journal of Political Research*, 17: 623–639.

Fields J. M., and H. Schuman 1976. 'Public Beliefs about the Beliefs of the Public', *Public Opinion Quarterly*, 40: 427–448.

Hirschman A.O. 1970. *Exit, Voice and Loyalty*, Cambridge, Mass.: Harvard University Press.

Hoffman, Wayne Lee and Terry N. Clark 1979. 'Citizen Preferences and Urban Policy Types', in John P. Blair and David Nachmias (eds), *Fiscal Retrenchment and Urban Policy*, London: Sage, pp. 85–106.

Kristensen, Ole P. 1980. 'The Logic of Political–Bureaucratic Decision-Making as a Cause of Government Growth', *European Journal of Political Research*, 8: 249–264.

Kristensen, Ole P. 1982. 'Voter Attitudes and Public Spending: Is There a Relationship?' *European Journal of Political Research*, 10: 35–52.

Lowi, Theodore 1969. *The End of Liberalism*, New York: W.W. Norton.

McIver, John P. and Elinor Ostrom 1976. 'Using Budget Pies to Reveal Preferences: Validation of a Survey Instrument', in Terry N. Clark (ed.), *Citizen Preferences and Urban Public Policy: Models, Measures, Uses*, London: Sage.

Ministère de l'Intérieur 1987. *Guide des ratios des communes de plus de 10000 habitants, 1985*, Paris: La Documentation Française.

Miranda, Rowan and Terry N. Clark 1988. 'Parties, Organized Groups and Political Leadership: Major Patterns in France and The US', paper prepared for distribution at the conference of the Fiscal Austerity and Urban Innovation Project, 2 September, Washington, DC: Brookings Institution.

Mouritzen, Poul Erik 1987. 'The Demanding Citizen: Driven by Policy, Self-Interest or Ideology?' *European Journal of Political Research*, 15: 417–435.

Mouritzen, Poul Erik and Kurt H. Nielsen 1988. *Handbook of Comparative Urban Fiscal Data*, Danish Data Archive, Odense Universitet.

Olson, Mancur 1982. *The Rise and Decline of Nations*, New Haven, Conn.: Yale University Press.

Peterson, Paul 1981. *City Limits*, Chicago: University of Chicago Press.

Rasinski, Kenneth A. 1989. 'The Effect of Question Wording on Public Support for Government Spending', *Public Opinion Quarterly*, 53: 388–394.

Samuelson, P.A. 1954. 'The Pure Theory of Public Expenditure', *Revue of Economics and Statistics*, 36: 387–389.

Smith, Tom W. 1987a. 'The Welfare State in a Cross-National Perspective', *Public Opinion Quarterly*, 51: 404–421.

Smith, Tom W. 1987b. 'That Which We Call Welfare By Any Other Name Would Smell Sweeter: An Analysis of the Impact of Question Wording on Response Patterns', *Public Opinion Quarterly*, 51: 75–83.

Wildavsky, Aaron 1980. *How to Limit Government Spending?* Berkeley: University of California Press.

Wilensky, Harold L. 1975. *The Welfare State and Equality: Structural and Ideological Roots of Public Expenditures*, Berkeley: University of California Press.

PART IV
RESPONSES TO
FISCAL STRESS

8

Fiscal Changes and Policy Responses:
A Comparison of Ten Countries

Norman Walzer, Warren Jones and Haakon Magnusson

Changes in local economic conditions and grants are ultimately reflected in policy outcomes: the degree of fiscal slack affects public policy. Previous chapters have examined some of the many factors that affect local decisions and make identification of policy strategies and responses difficult. None the less, it is important to examine city responses to changes in revenue sources.

Three main factors that are often supposed to affect policy outcomes were examined in earlier chapters. First is the degree of fiscal slack in each country. Major differences were reported (see Chapter 4). Finland and the Nordic countries, except possibly for Denmark, have had relatively little, if any, fiscal adversity. Finland, Norway, France and Sweden had more fiscal slack in 1985 than in 1978. Denmark, Italy and Canada had relatively little change. The UK experienced a substantial increase in stress during the study period.

Second was the relationship between intergovernmental aid received by cities and cutbacks in programs. Cities in countries where central governments reduced grants experienced more fiscal stress (see Chapter 4). Grants decreased, in real terms, in five of the ten countries examined (see Chapter 5); in Denmark, the UK and the USA the decrease was between 10 and 20 percent. In contrast, Norway and Finland received increases in grants of more than 40 percent. In only four countries did an increase in local revenue offset a decline in central government grants. Part of the explanation for these changes is that countries more dependent on grants are more vulnerable to reductions.

The third factor was 'color' of the party in control of city government. In Chapter 5 we were not able to detect a strong support for the party-political hypothesis, albeit the findings differed by country, clouding an interpretation of the effects. The number of observations was limited and the period of grant decline was short. In Chapter 7, on the other hand, we found a highly significant relationship between the mayor's spending preferences and party affiliation for two of the three countries investigated.

POLICY ISSUES

The policy responses to changes in fiscal conditions are the subject of this chapter. Strategies reported by city officials are examined in detail in the next chapter. Here we examine three questions.

First, does fiscal slack affect expenditure changes directly and separately from grants? Earlier chapters reported a high correlation between grant change and slack, which is attributable to the fact that grant change is a major component in our measure of fiscal slack. Other factors, however, may be as, or even more, important. For instance, in explanation of its spending changes, Norway experienced a major oil boom during the study period; and central government policies were unable to constrain the resulting expansions in local government programs. This prosperity reduced the relative importance of grants, and, even if grants had declined instead of increased, the cities might not have experienced significant strain. In this instance, expenditure growth did not rely solely on the grant system.

Second, to what extent are expenditure changes financed by increasing local taxes? Are cities in countries with an already relatively heavy property tax burden as willing to increase taxes to support expenditure as those relying more heavily on other revenue sources? The null hypothesis is that they are less likely to increase expenditure.

Third, which public services have had the greatest increase or decrease? Do these changes relate to government color? Earlier analyses classified spending preferences into allocation policies, urban development policies and municipal management (Chapter 7). An analysis of spending preferences found that mayors want to spend more in the USA, Finland, France and Norway, but less in Denmark; French mayors preferred a change in the distribution of expenditures. Unfortunately, detailed expenditure information is not available for each country, making tests of preferences and policy outcomes impossible.

Empirical tests were conducted at two levels, consistent with other analyses in this project. First, national data were examined. This information is highly aggregated and less directly comparable than city-specific data. Nevertheless, regression techniques were applied to these data. Second, when available, data on cities in each country were analyzed to capture the effects of city characteristics on expenditure changes. Multiple regression analysis was employed for five countries. In some instances, particularly Denmark, the variations among cities in expenditure changes are relatively small. This variable makes identifying the effects of each factor difficult.

Expenditure changes depend on at least four main factors, some of which have been mentioned but deserve further elaboration. First is availability of local resources. These resources often depend on prosperity in the national (or state) economy and on how well the local economy is integrated. The boom in oil prices, for instance, benefited cities in Norway where local public officials had the resources to institute new programs. In the USA, many states and cities had to seriously cut back on programs during the early 1980s. Some had barely reached pre-recession employment levels even late in the decade. Aggregate employment statistics do not always capture these differences, especially if manufacturing employment is replaced by lower-paying service industries. The public finance system may also adjust slowly. Services, for example, may not be taxed at the same rate as manufactured goods.

The second main factor is central government assistance in the form of categorical or general purpose grants. Categorical government aid is often used by central governments to increase local spending for specific services through a matching grant. The matching grant lowers the price of certain services to local governments. Directing local spending in this manner serves to divert spending from other services which, because of the categorical grants, become correspondingly more expensive. General purpose grants from a central government provide numerous options for local public services, not the least of which is a reduction in local tax effort. This was true in the USA, where state governments provided funds with the understanding that property tax relief would follow. Central government restrictions on such government support are an important consideration but are difficult to quantify in empirical analyses.

A third category of analysis is the perceptions of city officials about the fiscal situation or about citizens' preferences. Such perceptions may also determine spending practices. They may be subjective, or may be based on objective criteria. Information about perceptions of fiscal conditions was not available for all countries.

Finally, central government or constitutional mandates specify many programs that cities must provide. An increase in mandates, even without additional resources, can significantly increase city spending. Local public officials respond to mandates with varying programs, depending in part on available local fiscal resources.

NATIONAL POLICY COMPARISONS

Before investigating the propositions outlined above, broad comparisons of policy changes by country are provided (Mouritzen and Nielsen, 1988). These comparisons, which use national data for each country, are reported in Table 8.1. Most of the figures in the table show changes over the period 1978–85 in the form of indices with a value of 100 in 1978. A few of the figures measure the situation in percentages in 1985. (This is indicated at the end of the relevant row entries.) All indices involving monetary terms are based on constant prices (real terms). The countries are grouped into north European, south European and North American categories; cf. Chapter 2 above.

Socio-economic conditions
Wide variations in socio-economic characteristics are reported among country groups. Real income, on the average, increased most rapidly in north European countries, followed by south European and North American countries. Within each country group, however, major differences were reported. For instance, Finland reported an increase of 39.5 percent compared with a 0.2 percent decline in West Germany. The booming oil economy in Norway brought major inflation, an increase in the consumer price index of 79 percent. Even with substantially increased money incomes, purchasing power was eroded. Denmark and Sweden are the only other countries without a substantial real income increase during the study period. These countries also reported substantial inflation.

The unemployment rate, reflecting economic conditions, is also shown in the table. North European countries averaged 6.7 percent, substantially below the averages of 10.1 percent (south European) and 8.8 percent (North American) in other country groups. Within the north European countries, Britain, Denmark and West Germany reported unemployment rates higher than in other countries.

Comparison of income changes and unemployment rates suggests major differences in economic environments. In Denmark the relatively stable personal income occurred partially because of relatively high unemployment and, presumably, a surplus labor market. Inflation in Denmark was 80.1 percent during this period compared with 79.0 percent in Norway. Conversely, in Norway unemployment was only 2.6 percent during this period. While differences in inflation are not great, the additional effect of high unemployment in West Germany explains the relative real income decline. US cities reported relatively small increases in real incomes, largely as a result of double-digit inflation, but the unemployment rate was about average for all countries.

Italy clearly suffered the worst inflation, with a 168.7 percent price

Table 8.1 Trends in policy outcomes, 1978–85 (1978 = 100)

	North European						South European		North American		Average
	Denmark	Finland	Norway	Sweden	UK	Germany	France	Italy	Canada	USA	
Fiscal slack	100.7	137.6	131.3	108.6	92.7	103.0	121.3[1]	99.2[1]	102.3	97.0	109.4
Unemployment (1985 %)	9.3	5.0	2.6	2.8	11.3	9.3	10.2	10.1	10.5	7.2	7.8
Personal income	102.2	139.5	114.8	104.5	117.5	99.8	105.5	117.6	111.7	104.0	111.7
Tax base	109.2	133.1	123.8	105.2	108.3	103.0	113.2[1]	103.0	108.3	111.9	111.9
Tax effort	114.8	91.9	93.9	96.8	106.4	106.7	119.0[1]	159.6[1]	87.1	95.9	107.2
Total taxes	125.3	122.3	116.2	101.9	115.2	109.9	134.6[1]	161.3[1]	94.4	107.2	118.8
Reliance on tax revenue (1985 %)	48	39	56	41	29	35	35[1]	10[1]	37	42	41.3
Fees and user charges	121.4	115.9[1]	188.5	139.2	104.1	122.0[1]	123.5[1]	210.7[1]	–	124.7	138.9
Reliance on user fees and charges (1985 %)	22	15[1]	9	20	13	25[1]	9[1]	9[1]	10[1]	15	14.7
Total grants	90.1	148.2	147.8	114.9	85.0	102.9	128.3	99.0[1]	97.4	81.5	109.5
Current expenditure	122.1	132.9	131.4	127.2	97.7	117.9	130.1[1]	109.8[1]	–	110.9	120.0
Capital expenditure	49.3	99.2	74.5	74.1	71.1	80.5	110.6[1]	146.3[1]	–	107.4	90.3
Capital/total expenditure	44.4	78.8	61.6	63.5	76.3	75.2	90.3[1]	119.1[1]	–	97.4	78.5
Capital expenditure (1985 %)	6.8	16.2	11.3	12.7	13.2	21.7	35.3[1]	42.5[1]	23.3	15.7	19.9
Per capita municipal employees	121.5	119.0[1]	–	117.8	96.4	–	106.0[2]	102.9	–	88.3	107.4
Wage change	0.5	17.3	–	-4.3	2.7	–	22.5	17.5	–	10.9	9.6
Per capita municipal wages and salaries	122.0	136.3	133.0	113.5	99.1	108.7	128.5[1]	120.4[1]	–	99.2	117.9
Wages/currentexpenditure(1985%)	46.5	33.6	56.8	52.3	52.1	41.1	–	–	–	40.4	46.1
Consumer price index	180.1	180.7	179.0	187.6	189.3	132.8	198.2	268.7	171.9	164.9	185.3

[1] Data for 1978–85 are unavailable, so 1978–84 was used.   [2] Data for 1978–85 are unavailable, so 1978–86 was used.
Source: Mouritzen and Nielsen (1988: 26–52 and Table 2.1 of Country Index Tables for respective country).

increase between 1978 and 1985. Other countries with relatively high inflation include France (98.2 percent) and the UK (89.3 percent). One might expect that public expenditure increases, especially for traditional services, would be smaller in countries where real income is not increasing rapidly, or is even declining. Redistributive and social services, however, may increase during poor economic times.

A casual inspection of the data does not always support a direct relationship between higher expenditures and increases in income. Among north European countries, Norway and Finland reported an income growth of 14.8 and 39.5 percent, respectively. In both countries, however, we can observe a growth of current expenditures of about 32 percent. To complicate comparisons further, these countries had the two largest and almost similar grant increases. As a result, Norway substantially increased its reliance on tax revenues (56 percent), compared with a 39 percent increase in Finland.

Economic conditions are especially important to policy outcomes. When cities rely heavily on property or sales taxes with revenues determined by sales value, inflation can generate revenue increases without real income increases. This was a significant factor in the property tax limitation movement (Proposition 13) in California in 1978. Thus, cities in countries with high inflation may be better able to increase expenditure than countries where stagnant real income is explained by high unemployment. Residents are not always deceived by fiscal illusion, however, and may resist growth in government expenditure.

Per capita tax bases and reliance on tax revenues give the best description of the fiscal climate for financing municipal services. Countries reported major disparities in tax revenue reliance, with 56 percent in Norway and only 10 percent in Italy. North European countries, with 41.3 percent of revenues derived from taxes, are more similar to North American countries at 39.5 percent than are south European countries at 22.5 percent. Reliance on user fees and charges is comparable among the three country groupings. The north European countries averaged 17.3 percent of their revenue from user fees and charges, followed by North American countries at 12.5 percent and south European at 9 percent.

Per capita tax bases increased most in Finland, a country where cities derive approximately 39 percent of revenues from taxes. This tax base increase certainly could support the 32.9 percent current expenditure growth and the 36.3 percent wage and salary increase. Grants also increased 48.2 percent during this period.

Two of the ten countries (Italy and West Germany) reported minimal increases in real per capita tax base. The UK, with less than 10 percent increases, reduced current expenditure, reduced employees and lowered wages and salaries per capita. West Germany, with an essentially stable tax base (3 percent growth), increased current expenditure by 17.9

percent, provided wage increases of 8.7 percent and reduced capital expenditure by 19.5 percent over the period under study.

Policy outcomes
Policy outcomes are measured in many ways, but most often by expenditure and employment. Changes in expenditure are examined throughout this chapter. Because currencies vary by country, percentage changes are used. While the examination of differences in types of expenditures – allocation, development and management – was desirable, consistent information for all countries was not available.

Countries differ in real current expenditure growth of local governments. North and south European countries reported 21.5 and 19.9 percent increases, respectively, compared with 10.9 percent in North America in the period under study. Differences among countries within each group, however, are even greater. Among north European countries, Britain reported a 2.3 percent decline, compared with a 32.9 percent increase in Finland, the highest reported. France was second highest with a 30.1 percent increase. France is especially interesting because it reported only a 5.5 percent real income increase but had the third highest increase in grants. Grants were stable in Italy with only a slight decline; in addition, Italy, with a 17.6 percent income increase, reported only a 9.8 percent real current expenditure increase.

City officials, especially during short-term fiscal shortfalls or recessions, can postpone planned capital projects or reduce capital expenditure to protect current programs. There has been extensive research on how city officials choose between capital and current projects (Pagano, 1988). A comparison of capital expenditures among countries, however, is complicated by reporting variations. In some cities maintenance expenditure is reported as current expenditure, while in others it is shown as capital expenditure.

In studying capital expenditure, two aspects are especially important. First is the expenditure on planned capital construction. These projects do not have a clientele as such and may be delayed with less resistance. Second is repair to existing equipment. Continuing delay of maintenance expenditure presents a false picture of expenditure reductions. The costs continue to increase, especially for future replacements. Capital repair and maintenance projects are often not as visible and may not immediately affect residents as much as reducing critical services, such as police or fire protection.

City capital expenditure change differs among countries. South European cities were highest, with an average growth of 28.4 percent, while the north European countries were low, with a 25.2 percent decline. Italian cities had capital expenditure growth of 46.3 percent, the highest among all reporting countries. Since Italian cities reported relatively low current expenditure growth, they may be rebuilding capital facilities at the cost of current programs. This is especially interesting in light of the decline in real grants. Apparently, capital expansion is not financed by central government grants. Local tax effort in Italy, with increases of 59.6 percent, financed capital expansions during this period. This figure compares with an average 7.2 percent increase in all countries.

Six of the nine countries for which data are available reported declines in real capital expenditure. The two exceptions, in addition to Italy, are the USA (7.4 percent increase) and France (10.6 percent increase). Finnish cities were stable with a slight decline in real expenditure. Casual examination suggests that capital projects were reduced to preserve current operations, except in countries such as Norway and Finland, which did not experience serious fiscal stress.

Capital expenditure reductions also involve past capital expenditure. If, for instance, Nordic countries in relatively prosperous years were able to meet capital requirements, they could spend less on capital in later years. Discussions with local officials in Denmark supported this view, but the information needed to test this proposition was not available for all countries.

Commitment to capital spending, measured by proportion of total city expenditures, also affects spending. The cities averaged 19.9 percent of expenditure devoted to capital, but this information may be affected by differences in accounting and reporting systems. Cities known to be under fiscal stress, such as those in the UK and Italy, have very dissimilar commitments to capital spending. UK cities devoted an average of 13.2 percent to this purpose, while Italian cities spent 42.5 percent. To re-emphasize, past spending practices may be the important issue.

Capital expansion was also made at the cost of personal service expenditure in some cities. Major differences among countries were found in the growth of wages and salaries. Italy and France, with the highest increase in capital expenditure, also reported above-average growth in wages and salaries. Italy is particularly difficult to explain, with a 9.8 percent per capita current expenditure increase, but a per capita city wage and salary increase of 20.4 percent. France had one of the highest wage increases – 28.5 percent – but also reported the second highest capital expenditure increase. Tax effort in France increased by 19 percent during this period, second highest next to Italy. France is similar to Italy in its reliance on local resources to support expenditure growth.

Public officials in some countries can trade city employment for wage increases. Of the seven countries for which employment data exist, only two reported a decline in employment adjusted for population. The UK had a decline of 3.6 percent and the USA reported a decline of 11.7 percent. Both reported grant declines – 15 percent in the UK and 18.5 percent in the USA. These two countries also reduced real municipal wages by less than 1 percent. US cities clearly traded employment (11.7 percent decline) for capital expenditure (7.4 percent growth). Britain, on the other hand, reported a decrease of only 0.9 percent in per capita municipal wages and salaries, but a drop of 28.9 percent in capital expenditure.

DETERMINANTS OF CITY EXPENDITURE CHANGES

Pooled data analysis
While casual inspection of trends provides interesting comparisons of city responses to changes in fiscal resources, these comparisons are not sophisticated enough to capture more subtle changes in the countries under study. Better insight into variations among countries is provided by a correlation analysis of key variables. This analysis is based on pooled aggregate data, covering 10 countries for nine years.

Pooled regression analysis is necessary because of the small number of observations in each country or year. In conducting such an analysis, time differences must be separated from country differences. A recommended econometric approach is to insert dummy variables for each country to separate shifts arising from differences in country characteristics (Judge et al., 1982). Without this adjustment, interpretation of the relationships is ambiguous.

Because the dependent variable is the percentage change in expenditure, the independent variables are also expressed as percentage changes between 1977 and 1985 (Table 8.2). Each country was assigned a dummy variable to determine whether countries differ in spending changes after the effects of income changes, adjustments in grants and other factors had been considered.

In addition to income, grants and government 'color', other factors affect expenditure changes in Denmark, France and the USA (cf. the significant coefficients for these countries). Previous discussions of expenditure changes indicated possible reasons for these deviations from the average. In US cities, a noticeable shift to user charges occurred. This revenue source could explain some of the unanticipated revenue increases. Both France and Denmark had a substantial increase in tax effort.

Changes in local resources, as measured by personal income changes, are not an important predictor of spending changes. The regression coefficient was not significant at any commonly accepted level. Government 'color' was also not significant in the equation, but is related to changes in slack. When the regression equation is estimated with both slack and government 'color', the latter becomes significant with a negative sign, indicating that more conservative parties are associated with greater spending.

Table 8.2   *Pooled analysis*[1]

Independent variables	Beta coefficient	Level of significance
Personal income change	0.096	(0.311)
Total grants change	0.772	(0.000)
Government color[2]	–0.318	(0.120)
Dummy variables for country:		
Denmark	0.228	(0.048)
Norway	–0.056	(0.580)
Sweden	0.173	(0.114)
Finland	0.002	(0.987)
Germany	0.159	(0.153)
UK	0.012	(0.899)
France	0.167	(0.099)
Italy	0.157	(0.177)
Canada	–0.031	(0.718)
USA	0.233	(0.027)
Statistics		
R^2	0.5844	
R^2 Adj	0.4913	
F-ratio	6.2742	
Standard error	4.6394	
Degrees of freedom		
Regression	13	
Residual	58	

	Correlation			
	Change in total expenditure	Change in total grants	Government color	Change in personal income
Change in total expenditures	1.000			
Change in total grants	0.692	1.000		
Government color	0.089	0.080	1.000	
Change in personal income	0.233	0.195	0.101	1.000

Unit of analysis: yearly observations for countries (pooled data).
[1] Dependent variable is change in total expenditures.
[2] Government 'color' is such that: '1' is right-wing parties; '2' is right- and left-wing; '3' is left-wing parties.

Source: Mouritzen and Nielsen (1988: 26–52).

Statistically significant at 1 percent is change in grants. The ordinary regression coefficient (not in the table) shows that an increase of 1 percent in central government grants is associated, on average, with a 0.58 percent increase in expenditure, other factors considered. Thus, based on national pooled data, grants are substitutive rather than stimulative. Cities receiving an increase in grants tend to increase spending, but not by as much as the grant change. Cities may be using the grants to provide property tax relief, as noted in Table 8.1, especially in Finland, Norway, Sweden, Canada and the USA.

While grant changes are highly significant in the regression equation, multicollinearity among the independent variables is a problem. When fiscal slack is included in the estimated equation, the simple correlation between change in grants and change in fiscal slack is 0.86, which could easily affect the regression results. Change in fiscal capacity was highly correlated ($r = 0.61$) with percentage change in expenditures, but not with change in income or government color. Government color was not related to any of the other variables as indicated in the correlation matrix. Overall, the variables are associated with slightly less than one-half (49.1 percent) of the variation in spending changes. There are 58 degrees of freedom, and the F-ratio of 6.27 is highly significant.

This analysis is not based on city-specific information. A detailed understanding of factors important in spending changes must be examined at that level. City spending changes are included in the next section.

City-level analysis

Building on national comparisons, we now examine determinants of expenditure changes with city-level data. Relatively extensive literature exists on determinants of spending. Mouritzen (1989), for instance, examined spending changes as they relate to the election cycle. He found evidence of a political business cycle in local politics. Others have found strong relationships with previous spending patterns and socio-economic characteristics (Ladd and Yinger, 1989). Mouritzen found a general tendency for expenditure growth to be related to level of taxation and level of service at the beginning of the period. Elderly population, tax base, total population size, density, housing age and other factors are related to city spending. Present findings agree with those in earlier studies. Detailed information on expenditure changes is available in five countries, and regressions are estimated based on city data (Table 8.3). City characteristics measuring the expectations formulated earlier are included as independent variables.

Cross-country city-level analyses have several limitations. First, there is a lack of comparable data. While population, unemployment and fiscal slack are similar, other measures, such as the political party of the mayor, strength of council and form of government, are more difficult

Table 8.3 *Determinants of city spending changes*

Independent variables	Denmark	Norway	Sweden	Finland	USA
Population	–0.0930	0.0374	0.0746	–0.2043*	0.1035
	(0.1603)	(0.4562)	(0.2515)	(0.0510)	(0.1968)
Unemployment change	–0.1011	–0.0532	0.1588*	–0.1593	0.0475
	(0.1163)	(0.2891)	(0.0321)	(0.1506)	(0.5627)
Fiscal slack	0.2949*	0.2685*	0.2362*	0.3736*	0.3442*
	(0.0000)	(0.0000)	(0.0003)	(0.0012)	(0.0000)
Party of mayor	0.0692	0.0464	0.0131	–0.0956	–0.1889*
	(0.4055)	(0.5404)	(0.8908)	(0.4683)	(0.0219)
Socialist strength	–0.0471	–0.0547	–0.0620	0.0047	–0.0429
	(0.5975)	(0.4721)	(0.5514)	(0.9716)	(0.5927)
Statistics					
R^2	0.1382	0.0782	0.0667	0.2790	0.1737
R^2 Adj	0.1220	0.0657	0.0491	0.2275	0.1429
F-Ratio	8.5626	6.2582	3.7890	5.4171	5.6353
Standard error	4.6903	0.1445	0.9367	8.1304	0.2472
Degrees of freedom					
Regression	5	5	5	5	5
Residual	267	369	265	70	134

Unit of analysis: municipalities.
The results show the beta coefficients with the level of significance in parenthesis.
* Significant at 0.05 level.
Source: Mouritzen and Nielsen (1988: 26–52).

to compare. The second limitation is that institutions differ by country. Councils vary in responsibilities, mayors are elected for different terms, and arrangements between central and city governments are dissimilar. For these reasons, the results in Table 8.3 are preliminary and tentative.

Subsequent analyses are based on two tests. The first examines whether regression coefficients are statistically significant at the 5 percent level. The second examines whether the signs of the coefficients are similar across countries. Relationships presented earlier are examined – the effect of fiscal slack on expenditures and the importance of the mayor's political party in predicting expenditure changes. The degree of strength of the socialist party is also included to measure political power. This variable is more meaningful in Nordic countries than in the USA. The percentage of the council that is Democratic is included in the US regression.

To adjust for city socio-economic characteristics, population size and unemployment change are included as independent variables. These two variables are not only important in setting an agenda for city services, but are also directly comparable between cities.

The dependent variable is percentage expenditure change between 1980 and 1984, estimated in logs. This variable is affected by several factors. First, percentage changes are determined partly by expenditure in the base period. Cities that have met basic needs in the past may be able to reduce expenditure when confronted with the loss of local revenues or grants. This was noted earlier in Denmark. Second, cities may differ very little in percentage expenditure changes. When this is true, differentiating the effects of policy variables on expenditure changes is more difficult. This is true to some extent in the present study; cf. below.

The null hypothesis is that cities with increasing unemployment have less local ability to finance public services and, therefore, have smaller percentage expenditure increases. The only country in which unemployment change is significantly related to expenditure changes is Sweden. One explanation is that grants from central governments may be inversely related to local economic conditions so that unemployment increases trigger increases in aid from the central government. This arrangement stabilizes local public expenditure over the business cycle. Certainly, this was an objective of federal aid programs in the USA during the 1970s.

The sign of the unemployment coefficient for Swedish and US cities differs from other countries, but, because the relationship is not statistically different from zero in the USA, the sign has little meaning. Even though local public expenditure for programs associated with higher unemployment has increased rapidly in Sweden, it remains a relatively small part of the total city expenditure.

No statistically significant relationship was identified between population size and percentage expenditure changes in four of the five countries under examination. Finnish cities are the only ones for which percentage expenditure changes are related to population size, and in this case larger cities had smaller increases. The demands of residents for specific public services, even though related to city size, are difficult to quantify. Large cities, especially those with a decaying central city, may differ from suburbs, which have more resources and even lower service requirements. Large cities may especially need additional infrastructure (Ladd and Yinger, 1989).

Previous analyses examined correlates of fiscal slack,[1] especially the role of central government grants. In Table 8.3, fiscal slack is the only significant variable associated with percentage expenditure changes in the Nordic and North American cities. In each case, cities with more fiscal slack had higher percentage expenditure increases. The sign is intuitively correct.

The party of the mayor (dummy with 1 for leftist mayor), initially considered a key variable, is not significant in the four Nordic countries, but is significant for the USA. The strength of the socialist party is not important either. While unexpected, the lack of significance can be easily rationalized. Left and right political designations have blurred. Parties liberal on social programs may be fiscally conservative. The various possible political and spending strategies are examined elsewhere (Clark and Ferguson, 1983).

A contagion effect may also be working. Cities under the control of one political party may introduce a program or service, pressurizing other cities to institute similar programs or services. An example is Sweden, where in the 1980s socialists favored day-care programs for children and other programs for the aged. These were partially financed with central government grants. Populations in bourgeois suburban cities also needed such facilities. The programs spread, making the role of political parties difficult to differentiate in statistical analysis. The same effect has been found in Denmark concerning the development of day-care programs in the 1970s and the 1980s.

Grants are not included separately as an independent variable in Table 8.3, mainly because they are highly correlated with fiscal slack. When grants and slack are included in the same regressions, they are not statistically significant. While this finding might lead to a conclusion that grants do not function separately from fiscal slack, the high correlation with fiscal slack leaves such an interpretation risky.

The ability to predict percentage expenditure changes is relatively low. One explanation is the lack of differences in spending changes among Swedish and Danish cities. With such small variations, it is difficult to account for differences in spending changes. Even though the predictive ability is relatively low, the F-value is statistically significant.

SUMMARY AND CONCLUSIONS

Cities in the 10 countries examined in this chapter differ in the amount of fiscal slack encountered between 1978 and 1985. The Nordic countries of Finland, Norway and Sweden experienced relatively little fiscal austerity: in all instances there was more slack in 1985 than in 1979. Denmark experienced the tightest fiscal conditions, but even that does not compare with many other countries.

The UK, the USA and Italy experienced the greatest fiscal cutback among the countries in this study. The distress in the UK and the USA is attributable to declines in real central government grants. While, on average, grants increased by 9.5 percent over the period under study, they declined by 15 and 18.5 percent in the UK and the USA, respectively.

Changes in percentage-per-capita current spending increases, the measure of policy outcome in this chapter, also varied greatly among countries. Current expenditure increases ranged from 32.9 percent in Finland to –2.3 percent in the UK. Capital expenditures changed even more, ranging from a 50.7 percent decrease in Denmark to a 46.3 percent increase in Italy.

The relatively large decrease in capital spending in Denmark occurred even though local tax bases increased by 9.2 percent. Capital spending changes are especially hard to explain in Italy because the per capita tax base increased by only 3 percent. The major difference is that user fees increased 110.7 percent. This growth in user fees is misleading, however, because they represented a small portion of the city revenues initially. Thus, a relatively small increase reflected a large percentage increase.

This chapter examined three questions. Did fiscal slack, as measured in this analysis, affect city expenditures between 1980 and 1984? The answer clearly is affirmative. A strong and positive relationship was found between city spending increases and fiscal slack based on city-specific data for five countries. Fiscal slack is the only variable that was consistently significant in the regression analyses.

Are expenditure increases financed by local tax effort increases, or, alternatively, are reductions in expenditure prevented by increased local tax effort? In US cities, per capita real current spending increased by approximately 11 percent and capital spending increased by 7.4 percent. During the period, however, tax effort decreased by only 4.1 percent. Tax base increases replaced a portion of the decline in grants that could have supported additional expenditure.

Does government 'color' affect expenditure changes? Only in US cities is the political party of the mayor a significant explanatory variable. Experimentation with strength of the socialist party did not generate significant relationships. This lack of significance can be attributed partly to a blurring in party labels, and partly to the possibility that residents' demands for services outweigh party 'color' in determining expenditure changes.

Comparing policy outcomes among cities in ten countries is difficult. Governmental institutions and variations in arrangements for financing services complicate the analysis. Nevertheless, it appears from the analyses in this chapter that fiscal slack is an important consideration in policy outcome as measured by expenditures.

In this respect the comparative application of similar models in five different countries strongly confirms the general findings of the policy-output literature that public policy is determined by socio-economic characteristics (including the budget constraint: the degree of fiscal slack), rather than by political configuration (Fried, 1975; Sharpe and Newton, 1984).

NOTE

1 Fiscal slack was constructed by projecting revenues that would be received by local governments in the future at the same tax effort, including intergovernmental aid and comparing this amount with that actually received.

If the actual revenue exceeded the projected revenue, fiscal slack is said to exist. A positive number on the fiscal slack measure (opposite of fiscal stress) means less restrictive conditions facing local policy-makers. Including intergovernmental assistance in the fiscal slack measure complicates the analysis somewhat and makes differentiating the effects of fiscal slack and intergovernmental aid impossible.

REFERENCES

Clark, T.N. and L.C. Ferguson 1983. *City Money: Political Processes, Fiscal Strain, and Retrenchment*, New York: Columbia University Press.

Fried, R. 1975. 'Comparative Urban Policy Performance', in F. L. Greenstein and N. Polsby (eds), *The Handbook of Political Science*, vol. 6, Reading, Mass.: Addison-Wesley, pp. 305–379.

Judge, G.G., R.C. Hill, W. Griffiths, H. Lutkepohl and T.C. Lee 1982. *Introduction to the Theory and Practice of Econometrics*, New York: John Wiley.

Ladd, H. and J. Yinger 1989. *America's Aiding Cities: Fiscal Health and the Design of Urban Policy*, Baltimore: Johns Hopkins University Press.

Mouritzen, P.E. 1989. 'The Local Political Business Cycle', *Scandinavian Political Studies*, 12(1): 37–54.

Mouritzen, P.E. and K.H. Nielsen 1988. *Handbook of Comparative Urban Fiscal Data*, Odense: Danish Data Archives, Odense University.

Pagano, M.A. 1988. 'Fiscal Disruptions and City Response: Stability, Equilibrium, and City Capital Budgeting', *Urban Affairs Quarterly*, 24(1): 118–137.

Sharpe, L.J. and K. Newton 1984. *Does Politics Matter? The Determinants of Public Policy*, Oxford: Clarendon Press.

Walzer, N. and W. Jones 1988. 'Spending Trends in American Cities', presentation to Midwest Political Science Association, Chicago.

Wolman, H., E. Page and M. Reavley 1990. 'Mayors and Mayoral Careers', *Urban Affairs Quarterly*, 25(3): 500–513.

9

Choosing Fiscal Austerity Strategies

Norman Walzer, Warren Jones, Cecilia Bokenstrand and Haakon Magnusson

City officials are concerned about maintaining services and infrastructure when resources, local or intergovernmental, do not keep pace with inflation or are reduced in nominal terms. Their response to such fiscal cutbacks has been a topic of interest to local public officials and academics for many years (Burchell and Listokin, 1981; Levine, 1980). The worldwide recession in the early 1980s and the ensuing policy changes by central governments provided an opportunity to identify strategies selected by city officials in responding to fiscal changes.

The recession did not affect all cities alike, as has been demonstrated in earlier chapters. It is useful, nevertheless, to compare the responses in countries that faced significant fiscal setbacks with countries in which the economy was expanding. The Fiscal Austerity and Urban Innovation (FAUI) project is the most extensive worldwide data source on policy-maker responses to changes in city finance. The effort expended by FAUI in surveying cities attests to the major interest in city responses to adverse fiscal conditions. Two basic approaches exist for determining such responses. The first is to survey practitioners about which types of strategies they viewed as most important during a period of fiscal shifts; the second is to examine spending changes which represent policy responses to the fiscal changes.

This chapter examines perceptions of city officials about fiscal retrenchment strategies and whether these perceptions are consistent with reported fiscal data. Special attention is paid to trade-offs between two strategy groups: 'Increase user fees' versus 'Increase taxes', and 'Reduce capital expenditure' versus 'Reduce salaries and wages'.

ORGANIZATION OF THE CHAPTER

This chapter is based on strategies reported by municipal officials in the FAUI mail survey in industrialized countries. The strategy responses are country averages and the fiscal variables are country aggregates as reported in the *Handbook of Comparative Urban Fiscal Data* (Mouritzen and Nielsen, 1988).

The analysis is presented in two main sections. First, there is a review of major issues involved in selecting strategies, including a brief overview of the literature regarding strategy determinants. The FAUI project was initiated in 1982 and the academic literature is relatively small but increasing.

There then follows a comparison of policy outcomes with responses by chief administrative officers to questions about strategies used during fiscal austerity. Because policy outcomes have been reported in detail in a previous chapter, those responses are not repeated here.

FAUI survey information was collected between 1982 and 1987. The countries in this analysis were Denmark, Norway, Sweden, Finland, France, the UK and the USA because both survey and fiscal data are available for these countries. The Canadian survey included only officials in western Canada; the differences between survey and fiscal data coverage were regarded as too great to include Canada in subsequent discussions.

IMPORTANT DECISION-MAKING ISSUES

The fiscal austerity literature identifies several determinants of strategy selection. Political scientists stress political ideologies of city administrators, municipal employees and residents. Economists view economic conditions as particularly important, with fiscal stress and socio-economic conditions as dominant factors determining city fiscal policies. Sociologists study the effects of policies on population groups.

An examination of fiscal austerity responses should combine socio-economic factors with political attitudes and other city characteristics that determine the setting within which city decisions are made. Clark and Ferguson (1983) employ this approach and have set a standard for subsequent analyses of fiscal austerity. Their analyses combine socio-economic characteristics with political ideologies and attitudinal indices to explain decisions and policy selection. While studies in the fiscal austerity literature differ in methodology, many common variables are employed. The basic issues involved in the early literature are reviewed below as background for later analyses.

Political versus economic determinants
The public expenditure determinant literature began more than 30 years ago with the publication of independent works by Fabricant (1952) and Brazer (1959). Since that time, numerous efforts by economists, political scientists, and sociologists have identified correlates of public policy

outcomes (Dye, 1966; Bahl, 1969; Sharpe and Newton, 1984; and for reviews, Fried, 1975; Boyne, 1985).

These studies commonly employ multiple regression techniques with expenditure as the dependent variable and city characteristics (political, sociological and economic) as independent arguments. Often, independent variables are selected, a priori, without a generally accepted underlying theory of decision-making. The regression results can identify the relative importance of political and economic determinants but may leave researchers to rationalize the findings within commonly accepted views of how local governments function. Early studies by economists emphasized economic considerations, while later studies by other social scientists stressed political and attitudinal characteristics (cf. Fried, 1975).

Management techniques versus institutional setting
The motives of local public officials and the effects of these motives on resulting policy outcomes have been debated extensively in the political science and public finance literature. Some claim that bureaucrats seek to maximize budgets (Niskanen, 1971), while others argue that incremental budgeting models determine spending levels (Davis et al., 1966a, b). Partisan politics, government reform and application of business techniques to city management are also widely discussed in the fiscal austerity and urban innovation literature (Clark and Ferguson, 1983; Baldersheim and Hovik, 1987). Characteristics and management styles of city officials seem critical but are difficult to quantify and evaluate.

Innovative management practices also have been addressed in the fiscal austerity literature (Appleton, 1989; Skovsgaard, 1985). One hypothesis in this line of research asks whether the timing of fiscal strain affects the types of strategies adopted after other factors such as political ideology have been considered.

PROFILE OF THE SEVEN COUNTRIES

Following earlier chapters, we present countries in three broad groups, based on the importance of local government and other characteristics (cf. Chapter 2). These groups are then used to examine responses toward fiscal changes. Countries with limited fiscal stress may respond differently from those experiencing greater strain. The slack measure for a particular year is the ratio of actual revenue to projected revenues, assuming that tax rates are the same as in the base year. (Cf. Chapter 3 on the definition of slack.) When comparing relative changes, a larger number represents more slack.

The *north European countries* include Denmark, Norway, Sweden and Finland, countries that, in general, have had relatively high fiscal slack. In some ways, the UK (represented here with survey data from England and

Table 9.1   *Unemployment rates and fiscal slack*

	Denmark	Norway	Sweden	Finland	France	UK	USA
Year of survey	1983	1985/86	1986	1985	1985	1987	1983/84
Unemployment rate[1]							
High	10.5	3.3	3.5	7.3	9.9	11.5	9.7
	(1983)	(1983)	(1983)	(1978)	(1984)	(1986)	(1982)
Low	6.1	1.1	2.1	4.7	8.2	5.1	5.8
	(1979)	(1978)	(1979)	(1980)	(1982)	(1979)	(1979)
Fiscal slack by year (base year 1978)							
1979	101.8	103.6	104.8	104.7	104.4	101.9	98.5
1980	104.8	109.5	103.4	107.7	105.9	97.2	96.0
1981	105.5	110.9	105.5	109.7	109.3	89.3	96.1
1982	108.1	114.9	105.4	114.4	112.4	86.7	94.1
1983	106.9	116.4	105.3	123.0	114.0	90.8	91.5
1984	105.7	123.4	108.1	131.0	121.3	91.0	95.2
1985	100.7	131.3	108.6	137.6	–	92.7	97.0
1986	100.4	–	111.0	–	–	–	–

Unit of analysis: countries.
[1] Unemployment from national census sources; fiscal slack calculated as explained in Chapter 3.
Source: Mouritzen and Nielsen (1988: 39 and 48).

Wales) also fits this group, but with much less slack (Table 9.1). Finland and Norway clearly are in the best fiscal position; Sweden and Denmark are next, followed by Britain. Britain was the most fiscally strained of all countries in the survey. Rankings from low to high unemployment rates follow a similar pattern.

The level of public services varies by country and by the way responsibility for services is shared between central and local governments. The northern European group relies heavily on city governments to provide services, as shown by the high proportion of total public expenditure represented by cities. In 1984, municipal expenditure as a percentage of total public expenditure was 44.1 percent in Finland, 29.7 percent in the UK, 27.9 percent in Sweden, 27.3 percent in Denmark and 24.8 percent in Norway (Mouritzen and Nielsen, 1988: 35). In contrast, the percentages in Italy, France and the USA were 14.4, 13.2 and 11.1, respectively.

The group of *south European countries* is represented by France, which has a greater differentiation between private and public sectors than north European countries. French cities reported less fiscal slack in the years prior to the FAUI survey than Finland and Norway, but were better off than the cities in Sweden, Denmark and the UK. The level of French unemployment in the year of the survey was similar to that in Denmark.

The USA represents the *North American* group. American cities faced a major recession in 1981–82 and had a conservative federal government throughout the 1980s. Federal support for local governments as a proportion of GNP declined from the late 1970s onwards. The recession was particularly severe for cities in the North-east and Midwest, although the North-east responded more positively to the recession. Cities in the USA and the UK both reported an increase in fiscal slack by 1984, but they are the only countries in which cities had less fiscal slack at the end of the period than at the beginning.

In Appendix 9A we have drawn a more elaborate profile of each country (including also West Germany, Italy and Canada), which involves socio-economic and policy outcome trends over the 1978–84 period. In the analysis below we refer to these figures when comparing policy outcomes with reported strategy use.

COMPARISONS OF STRATEGIES

Reported strategies are grouped into meaningful classifications to compare strategy responses with policy outcomes (cf. Table 9.2). Many classification systems could be devised using statistical analyses. Because of the difficulties associated with making comparisons between countries, the strategy groupings in this chapter are based on extensive discussions with FAUI researchers in the countries involved. Special care was taken to insure that the strategies reflect comparable responses by country.

In some instances, a single strategy impacts on a single fiscal outcome. An example is 'Increase user fees', which directly impacts on user fees. Often, though, a single fiscal variable is impacted by several strategies. In these cases the average value of these strategies is compared with the fiscal outcome.

This section compares reported use of the strategies presented in Table 9.2 by country. Not all strategies are equally appropriate in each country, however. For example, in some countries national wage negotiations may prohibit local authorities from reducing wages or employment. Strategy groups not comparable with policy outcomes are noted below.

Strategy selections are not mutually exclusive. City officials can combine revenue increases with expenditure reductions or with contracting for services. Questionnaires differed by country, but in most instances asked for the importance of a strategy within the previous five years. The Norwegian questionnaire covered three years prior to the survey. Sweden asked only for strategy selections in the survey year; for this reason, extra care has been exercised in interpreting Swedish results.

There is growing evidence that differences in methods of selecting and grouping strategies affect the ability to explain strategy selections statistically (Clark and Walter, 1986). Indeed, initial attempts to explain

Table 9.2 *Strategy classifications*

Increasing revenues
A. Increase user fees
B. Increase taxes
C. Other revenue-raising options
 1. Obtain new local revenues
 2. Add intergovernmental revenues
 3. Lower surpluses
 4. Sell assets
 5. Increase short-term borrowing
 6. Increase long-term borrowing

Expenditure strategies
A. Reduce expenditures without personnel actions
 1. Defer payments to next year
 2. Make cuts in all department
 3. Make cuts in least effective departments
 4. Reduce administrative expenditures
 5. Reduce expenditures for supplies, equipment, and travel
 6. Reduce services funded by own revenues
 7. Reduce services funded by intergovernmental revenues
 8. Eliminate programs
 9. Reduce capital expenditure
 10. Keep expenditure increases below inflation
 11. Defer maintenance of capital stock
 12. Control construction to limit population growth
B. Reduce expenditures through personnel actions
 1. Lay off personnel
 2. Reduce employee compensation
 3. Freeze wages and salaries
 4. Freeze hiring
 5. Reduce workforce through attrition
 6. Initiate early retirements
 7. Reduce overtime

Management strategies/productivity increases
 1. Improve productivity through better management
 2. Improve productivity by labor-saving techniques

Delegating responsibilities through contracting out or co-operation
 1. Shift responsibilities to other units of government
 2. Contract out services to other units
 3. Contract out to private sector
 4. Introduce purchasing agreements

strategy groups statistically (at the city level of analysis) may have been limited by the fact that absolute strategy values rather than their relative importance were used. When a strategy index is low in comparison to the average of all strategies, one might expect limited use. In such cases, fiscal outcomes will not be heavily influenced. This study considers the importance of a strategy or strategy group in relation to the average index score for all strategies in each country. In each table we report the absolute value as well as the average across all strategies in each country.

The absolute index score attached to each strategy in the tables below may range from 0 to 100. A score of 0 for a country indicates that no respondents from that country reported the use of that particular strategy; a score of 100 shows that all respondents reported that they had used the strategy and had found it to be very (or most) important. In Appendix 9A the exact procedure for each country is discussed in more detail.

REVENUE STRATEGIES

Increase user charges
Increased reliance on user charges has gained popularity in financing public services, especially by conservative administrations. The main rationale for user fees is that residents who benefit directly from services should pay. In this sense, user fees introduce a sense of fairness and allow local officials to avoid, to some degree, the political risks of increasing tax rates.

At the same time, fees are regressive if no adjustment is made for differences in income. One might expect socialist or liberal governments to resist burdening relatively poor residents. User charges and fees may be accompanied by a system of transfer payments to low-income residents to offset regressivity.

Contrary to expectations, given its large increase in user fees, Sweden reported user fees as least important (Table 9.3). In other countries this strategy was more important than the overall average, and user fees increased in every country. Norway had the greatest increase, 77.9 percent, and the greatest differential between this strategy's value and the overall average. User fees comprised a small percentage of total revenue in Norway in 1978, and, even with this dramatic increase, only 9 percent of total revenues were obtained from this source. (Cf. 'Reliance on user fee' in the figures in Appendix 9B.)

Despite discussions and federal emphasis, reliance on user charges

Table 9.3   *Increasing user fees as a revenue strategy*

Indices	Denmark	Norway	Sweden	Finland	France	UK	USA
Strategy index	51	60	16	49	46	45	58
Average, all strategies	42	24	19	41	36	36	31

Source: Mouritzen and Nielsen (1988: 12).

in the USA increased only from 12 to 14 percent of total revenues between 1978 and 1984. This is particularly interesting since user charges have received widespread attention in the USA. For example, a major requirement of federal funding for waste-water treatment is a system of user charges.

Denmark, Finland and the UK reported that user charges were especially important, and this strategy index is approximately 20 percent higher than the average for all strategies, but user fees increased only by 10.5 and 15.9 percent in Denmark and Finland, respectively. One plausible explanation is that the relative prosperity of these countries meant that major increases in revenues were not needed.

Only in Sweden is the fiscal policy outcome clearly inconsistent with the strategy responses. Despite the reported relative unimportance of user fees, they increased by 33.7 percent between 1978 and 1984. One apparent explanation for this poor correspondence is that the Swedish survey did not ask about past strategy decisions; that is, the period 1978–84 is outside the period covered by the strategy responses. At the time of the survey, user charges may already have been increased substantially and were no longer regarded as an important source of additional revenue.

Increase revenues with taxes

Tax increases are a second obvious revenue-raising possibility. In this analysis, *tax effort* is used as a policy response measure (Table 9.4). Tax effort is the ratio of tax revenue to tax base; if tax effort increases, an increase in average tax rates can be assumed. Because detailed information is not available for specific taxes, the importance of tax increases overall was used. With a combined tax measure, it is possible that some taxes increased while others decreased, leaving the overall tax effort the same.

Countries differ widely in the relative importance of the type of municipal taxes employed (cf. Table 2.3 above). Most north European countries rely heavily on local income taxes, while the USA and UK depend more heavily on property taxes. French cities are in between, relying more heavily on the income tax but not as heavily on property taxes.

Countries in which the tax burden or effort was high initially may face greater political opposition when raising taxes. Norway showed some slight inclination to raise taxes, and its initial tax effort was relatively low compared with surrounding countries (Table 9.5). Sweden and Finland, with relatively high tax effort and fiscal slack initially, avoided tax increases. Total municipal tax receipts increased in all countries, but tax effort increased only in Denmark, the UK and France. Tax effort decreased in Finland, Norway, Sweden and the USA. The decrease in tax effort cannot be attributed solely to the availability of other revenues, because Finland, Sweden and the USA differ considerably in

Table 9.4   *Index of municipal taxes per capita, tax base per capita and tax effort*

Indices	Denmark	Norway	Sweden	Finland	France	UK	USA
Municipal taxes per capita							
1978	100.0	100.0	100.0	100.0	100.0	100.0	100.0
1979	109.6	99.5	99.8	101.9	103.5	103.9	99.7
1980	114.1	99.3	97.1	104.1	105.9	106.1	98.2
1981	112.1	104.9	99.9	108.6	110.6	114.2	98.4
1982	112.2	107.1	99.2	108.8	116.9	119.4	99.8
1983	116.8	106.7	106.2	111.3	123.1	113.4	100.9
1984	119.5	108.4	102.3	117.3	134.6	116.5	104.1
Tax base per capita							
1978	100.0	100.0	100.0	100.0	100.0	100.0	100.0
1979	101.3	103.2	102.7	104.4	102.4	103.5	102.8
1980	102.6	108.1	100.8	107.1	101.7	98.6	102.3
1981	101.9	109.5	102.7	109.1	101.6	95.3	104.4
1982	103.2	111.4	101.9	113.0	103.5	96.5	103.9
1983	104.6	111.9	101.2	122.7	103.5	97.1	103.3
1984	105.5	115.4	104.1	137.5	113.2	98.9	108.2
Tax effort							
1978	18.09	13.39	20.70	15.85	11.66	4.03	2.65
1979	19.57	12.91	20.12	15.47	11.79	4.04	2.57
1980	20.10	12.31	19.95	15.41	12.15	4.33	2.54
1981	19.90	12.83	20.13	15.78	12.69	4.83	2.50
1982	19.66	12.88	20.14	15.27	13.18	4.98	2.54
1983	20.18	12.77	21.72	14.28	13.87	4.71	2.59
1984	20.48	12.58	20.34	14.58	13.88	4.75	2.55

Source: Mouritzen and Nielsen (1988: 27, 29, and 30).

fiscal slack. Only in Britain was there a tax base decrease (1.1 percent) during the period.

Denmark, the UK and France reported tax rate increases (tax effort increases), and all of these countries seemed to rely on the 'Increase tax' strategy. The USA is the only country for which the strategy index indicated that this strategy was important, yet overall tax rates did not

Table 9.5   *Increasing taxes as a revenue strategy*

Indices	Denmark	Norway	Sweden	Finland	France	UK	USA
Strategy index	72	27	14	29	67	63	47
Average all strategies	41.5	24.3	19.3	40.5	41.8	36.1	30.5

Source: Mouritzen and Nielsen (1988: 12).

increase in US cities. Finland, Norway and Sweden, which experienced relatively low levels of fiscal strain and unemployment, may not have needed tax increases to maintain services.

Relatively low levels of fiscal slack are likely contributors to tax increases. Denmark, the UK, France and the USA faced fiscal cutbacks with unemployment, and among these countries only US cities did not raise overall tax rates. The 1978 tax effort of 2.65 percent was the lowest of any country. At the same time, however, Denmark raised tax rates even with the second highest tax effort, 18.09 percent. Denmark's reliance on tax increases is partially explained by the fact that in 1982 a series of reductions in central government grants to local governments was implemented.

The UK, second lowest in tax effort, increased tax effort 17.9 percent (see Figure 9.5 below), but the tax base had declined 1.1 percent between 1978 and 1984. In this case, high fiscal strain, high unemployment and low tax effort brought higher tax rates. The USA is the only country in which outcomes do not fit the expected pattern: despite relatively high unemployment, low tax effort and a reported importance of raising taxes, tax collections declined.

Tax revenue sources do not completely explain differences between strategy selection and fiscal outcomes. The UK and the USA are similar in tax effort, unemployment and reliance on property taxes. However, while the UK municipalities raised tax rates, as expected, US cities did not, further documenting the case for inconsistency in the US strategy responses.

Other revenue-raising strategies

City officials balance budgets with revenue sources other than taxes and user charges. These options can be grouped into three basic types: 'Obtain new local revenues', 'Lower surpluses or sell assets' and 'Increase borrowing'. The 'Other revenue-raising options' strategy group could not be compared directly with policy outcomes. Variables necessary for the comparison either were not available or could not be aggregated in a meaningful way. Since each strategy is fairly distinct and could differ in appeal to city administrators with alternate philosophies and views about public financing, they are discussed separately.

The importance of 'Other revenue-raising strategies' as a group dif-fered among countries (Table 9.6). France and the UK reported a high degree of reliance on these strategies but apparently used different approaches. British cities relied on lowering surpluses, selling assets and seeking additional grants; French cities turned to long-term borrowing and generating new local revenues. Examining these reported strategies in a vacuum is hazardous, because during this time changes in central government policies can open or close funding opportunities available

Table 9.6   *Other revenue-raising strategies*

Strategies	Denmark	Norway	Sweden	Finland	France	UK	USA
New local revenue	–	24	2	63	42	38	54
Additional grants	8	30	–	–	39	55	34
Lower surpluses	71	39	–	–	–	69	44
Sell some assets	22	35	46	14	32	57	19
Borrow short-term	–	16	37	12	35	17	19
Borrow long-term	35	45	–	39	52	35	21
Average, other revenues	34	31	28	32	40	45	32
Average, all strategies	41.5	24.3	19.3	40.5	41.8	36.1	30.5

Source: Mouritzen and Nielsen (1988: 12).

to cities. Past spending patterns on infrastructure, for example, may also affect French long-term borrowing strategies.

Finland and Norway, which were relatively prosperous during this period, relied more heavily on long-term borrowing. Increasing debt may have been less of a response to fiscal cutbacks than expanding the infrastructure. This is especially the case in Finland, which also reported that generating new local revenues was very important. Denmark, facing cutbacks in central government grants, lowered surpluses and was similar to the UK and the USA.

The effects of the conservative federal government strategies in the USA are clear in the relative reliance on new local revenue sources reported by US cities. While additional grants also were reported as relatively important, they probably were from the state, rather than the federal, government. Lowering surpluses was also reported as an important strategy, which makes sense in light of other events underway during that period.

Sweden is the only country that regarded 'Increase short-term borrowing' as important, ranking it second among other revenue-raising strategies. Cities in all other countries avoided this strategy. The Danish questionnaire omitted it.

Separating long-term and short-term debt is difficult with existing data, but the debt changes are interesting. Many countries during this period reduced debt levels. Denmark, in particular, is interesting because the outstanding debt in 1986 was only 55.3 percent of the 1978 level. Debt reductions may mean that extensive capital projects are no longer needed or are financed with other sources, such as grants. Indeed, only Finland and France increased capital expenditure between 1978 and 1984. In Denmark capital expenditure in 1984 was only 40.7 percent of the 1978 level. Also, grants in Denmark were 6.0 percent higher in 1984 than in 1978 and had been consistently above the 1978 level. The dramatic drop in capital expenditure in Denmark cannot be explained by high fiscal strain; in this case, more than adequate public capital was available and expenditure priorities were rearranged.

EXPENDITURE STRATEGIES

Expenditure reduction strategies

Reducing expenditure is an alternative to raising revenues. Denmark reported this strategy as very important (Table 9.7). There, the reductions were triggered by the completion of capital projects (cf. the dramatic reduction in capital expenditure of about 60 percent in Figure 9.1 below), but also by reductions in central government grants. Finland and France also reported expenditure reductions as an important strategy

Table 9.7   *Reducing expenditures*

Strategies[1]	Denmark	Norway	Sweden	Finland	France	UK	USA
Defer payments cuts	–	15	–	32	20	42	18
Cuts							
All departments	25	14	20	77	44	48	38
Least efficient	–	67	22	65	38	33	25
Administrative	–	15	15	60	67	35	32
Supplies, travel,							
and equipment	–	31	11	45	–	25	39
Services funded with							
Own revenues	44	14	–	27	–	–	23
Grants	36	9	–	13	–	–	22
Eliminate programs	–	10	5	34	9	27	26
Capital	81	27	21	–	65	42	39
Increase less than							
inflation	–	26	27	52	–	36	37
Maintenance of							
capital	59	41	11	–	32	36	22
Limit growth	–	–	–	–	–	–	9
Averages							
Without personnel							
actions	49	24	17	45	39	36	28
With personnel							
actions	39	11	12	31	38	25	28
All expenditure							
strategies	46	20	15	40	39	31	28
All strategies	41.5	24.3	19.3	40.5	41.8	36.1	30.5

[1] For detailed description of strategies, see Table 9.2.
Source: Mouritzen and Nielsen (1988: 12).

with across-all-department cuts and administrative or capital reductions as particularly important. Norway, Sweden, the USA and the UK avoided expenditure reductions as a group. In Norway and Sweden such expenditure reductions may not have been needed. In Britain and the USA, because of greater fiscal pressure, a conscious decision about not reducing expenditure was needed.

Expenditure reductions take many forms. Often, the simplest approach is to postpone capital projects or new programs having little political support. Delaying equipment purchases or postponing capital improvements is only a temporary solution, however. There is clear evidence that during this period US cities reduced capital relative to current expenditure, presumably as an alternative to reducing employment (Walzer and Jones, 1988): between 1980 and 1984, capital expenditure decreased by 14.9 percent, consistent with strategy responses. After 1984, capital expenditure rebounded.

The French outcome is not consistent with the reported importance of this strategy, but Denmark and Britain experienced substantial declines in capital expenditure – 59.3 and 16.6 percent, respectively – consistent with the reported importance. While Norway and Sweden did not cite this as an important strategy, capital expenditure declined by 25.1 percent in Norway and 26.1 percent in Sweden. With the exception of France, where city officials indicated relatively heavy usage, the outcomes are consistent with reported importance.

The 59.3 percent decline in capital expenditure in Denmark during the period 1978–84 is striking. There is clear evidence that this reflects the belief by local officials that the capital stock was more than adequate.

Unlike the 'Reduce capital expenditure' strategy, little agreement is found among countries that any other of the individual strategies were important. However, there was some agreement about strategies that were not important (Table 9.6). Only Britain reported 'Defer payments' as relatively important. The index for 'Reducing services funded by intergovernmental revenues' was below the country average in all cases. 'Eliminating programs' was even less popular.

Cities in Denmark and the USA experienced little expenditure growth over the 1978–84 period – 5.4 and 5.3 percent respectively. (On increases in total municipal expenditure, cf. Mouritzen and Nielsen, 1988: 26.) These results are not inconsistent with the modest usage of expenditure strategies indicated in the surveys. Total expenditure in the UK declined by 4.7 percent during the period. Given that British cities did not strongly favor this strategy, the decrease, although small, does not seem entirely consistent with their strategy responses. Expenditures in Sweden, Norway, Finland and France increased by 17.0, 12.7, 22.0 and 22.5 percent, respectively. The expenditure increases in Sweden and Norway are not inconsistent with the low usage rate reported.

Personnel reduction strategies

Expenditure can be reduced in several ways through personnel actions, ranging from reducing overtime to terminations or lay-offs. Personnel strategies are especially important because wages and salaries are often the largest expenditure item in a municipal budget. Not all personnel strategies are viable in each country; in Denmark, for instance, 'Reducing employee compensation' has little meaning as a local policy alternative. Because a more detailed comparison of personnel strategies is provided in the following chapter by Hoffmann-Martinot, only a limited discussion of personnel strategies follows.

Most countries did not report intensive use of strategies involving personnel reductions (Table 9.7). The average index value for this group of strategies ('Averages with personnel actions') is well below the overall country scores in Norway, and somewhat below the average in the UK, Finland and Sweden. In the remaining countries, the average for this group is slightly below the overall average.

For comparison with strategies involving personnel actions, the change in total wages and salaries during the period is used, since it includes the effects of both changes in the number of employees and changes in employee compensation.

Norway, Sweden and Finland reported a low relative importance for this strategy group, and it is not surprising that salaries and wages increased in these countries (cf. 'Municipal wages' in the figures in Appendix 9B). Sweden experienced the lowest growth rate in wages and salaries, 12.8 percent. Norway and Finland had increases of 28.2 and 31.1 percent, respectively. The UK avoided this strategy group, and wages and salaries declined slightly, 0.7 percent. Denmark and the USA, which reported average usage relative to other strategies, show markedly different outcomes. In Denmark wages and salaries increased by 19.6 percent; in the USA total wages and salaries declined by 4.6 percent over the period 1978–84.

Norway, Sweden, Finland and France are consistent in strategy responses and policy actions. In the UK and the USA the case for consistency is weaker. Only in Denmark is clear inconsistency found between the strategy responses and policy actions.

MANAGEMENT, PRODUCTIVITY AND
DELEGATION STRATEGIES

Management/productivity improvements

The responses for the two productivity strategies included in the survey indicate the popularity of such strategies (Table 9.8). In six of the countries, the average index for the productivity strategies is above the

Table 9.8   *Productivity improvements*

Strategies	Denmark	Norway	Sweden	Finland	France	UK	USA
Management	71	42	58	67	66	54	54
Labor-saving	72	45	40	64	31	47	45
Averages for productivity strategies	–	44	49	66	49	51	50
Average for all strategies	41.5	24.3	19.3	40.5	41.8	36.1	30.5

Source: Mouritzen and Nielsen (1988: 13).

overall country average for all strategies. Only in France is the index for this strategy group just moderately above the country average; this is because only in France is the labor-saving strategy index below the country average. In all countries, the management strategy index is well above the country average index. While the management strategy index is above the labor-saving index for all countries but Norway, only in France is the management strategy clearly preferred to the labor-saving strategy.

A comparison of policy outcomes with the stated importance of these strategies is not possible because a direct relationship between the strategy and the policy outcome cannot be observed with the aggregate data available. For instance, improved productivity through better management or labor-saving techniques could result in a lower expenditure for the same level of service. However, the reduction in costs could result in an increase in the level of services. In either case, the cost per unit of service would decrease so that productivity improvements could be identified. However, detailed comparisons of the per-unit cost of municipal services are not available.

Delegating, contracting out and co-operation strategies
Only France and Finland reported serious use of this group of strategies (Table 9.9). However, in only a few instances is there a strategy with an index value substantially higher than the overall average strategy index. Sweden reported delegating responsibilities to other units of government as highly important. Entering into purchasing agreements with other units was reported as an important strategy among French cities.

As with the previous strategy group, aggregate data are inadequate to address the issue of whether these strategy responses are reflected in policy outcomes. Information is required on changes in functions performed by municipalities, changes in the number and size of contracts and changes in the size and number of purchasing agreements.

Table 9.9   *Delegating out responsibilities/contracting out/co-operation*

Strategies	Denmark	Norway	Sweden	Finland	France	UK	USA
Delegate to other units	–	12	31	–	–	–	16
Contract out to other units	8	28	7	–	–	0	15
Contract out to private sector	–	9	6	39	22	30	27
Purchasing agreements	33	3	15	43	56	39	25
Average for delegate/co-operation strategies	21	13	15	41	39	23	20
Average for all strategies	41.5	24.3	19.3	40.5	41.8	36.1	30.5

Source: Mouritzen and Nielsen (1988: 12).

TRADE-OFFS BETWEEN EXPENDITURE CATEGORIES

Increasing user fees versus increasing taxes

In responding to cutbacks, city officials can adjust user fees along with taxes. The strategy values for 'Increase user fees' and 'Increase taxes' were compared and the strategy with the highest value is regarded as one that local officials consider most important. Local officials in Denmark, France and the UK seem to have increased taxes as compared with user fees. In Denmark and Britain, tax rates increased more than user fees, consistent with the reported strategies. User fees and tax rates increased nearly the same in France, which is not consistent with a reported emphasis on tax rates.

Greater emphasis on user fees was reported in Norway, Finland and the USA. Observed changes in user fees and tax rates were consistent with the survey responses in all three cases. In Sweden the strategies were rated as being of nearly equal importance, with a slight tendency to favor user charges. The observed policy outcome was substantially higher user fees and lower tax rates. The fact that the Swedish questionnaire covered a period beyond the available data may explain this strong inconsistency.

Trade-offs between capital and personnel

Prioritizing city expenditures is an integral part of a policy response. Selective expenditure reductions require choices among services. Of special interest is the choice between capital projects and wages and salaries.

In all countries, the importance of the 'Reduce capital expenditure' strategy is higher than the 'Personnel actions' strategy group. The difference between these two categories is smallest in France and the USA. Given these strategy responses, capital expenditure should have decreased more (or increased more slowly) than total wages and salaries. With the USA as a notable exception, this indeed was the experience between 1978 and 1984. In the USA, salaries and wages declined by 4.6 percent while capital expenditures declined only by 0.3 percent. In all countries except the USA, greater emphasis on reducing capital expenditure relative to reducing total wages and salaries was found both in the strategy responses and in the policy outcomes.

GENERAL OBSERVATIONS

This chapter represents a first step in explaining fiscal austerity responses by city officials. Much more work is needed before a definitive model can be developed and fully tested. This analysis, however, offers several interesting findings.

Substantial consistency is found among strategies and fiscal policy outcomes for seven countries and six strategy groups (Table 9.10). In 48 such comparisons, 32 fiscal outcomes were consistent with reported

Table 9.10   *Consistencies between reported strategies and policy actions*

	Denmark	Norway	Sweden	Finland	France	UK	USA
Increase user fees	Yes	Yes	No	Yes	Yes	Yes?	Yes
Increase tax revenues	Yes	No?	Yes	Yes	Yes	Yes	No
Reduce capital expenditures	Yes	Yes?	Yes?	na	No	Yes	Yes
Reduce salaries and wages	No	Yes	Yes	Yes	Yes	Yes?	Yes?
User fees relative to taxes	Yes	Yes	No?	Yes	No	Yes	Yes
Capital relative to salaries and wages	Yes	Yes	Yes	Yes	Yes	Yes	No
Survey response rate (%)	77.5	82.4	80.3	85.2	46.2	65.5	49.3

Yes   Fiscal outcomes and reported strategies consistent.
No   Fiscal outcomes and reported strategies inconsistent.
Yes? Fiscal outcomes and strategies marginally consistent.
No?   Fiscal outcomes and strategies marginally inconsistent.
na   Strategy question not included in survey.

strategy usage; only 6 were judged clearly inconsistent; marginal consistency exists in 5 cases, and marginal inconsistency was found in 5.

It is not surprising to find inconsistencies between strategy responses and policy outcomes. The reported strategies are for a subset of cities within each country while the fiscal outcomes are based on aggregate data for the entire country. Sampling differences between survey and non-survey cities may occur. This is less likely to explain inconsistencies in Norway, Sweden and Finland, where survey response rates were 82.4, 80.3 and 85.2 percent, respectively. For France, the UK and the USA, with survey response rates of 46.2, 65.5 and 49.3 percent, respectively, the case for attributing differences in reported strategies and fiscal outcomes to differences between the sample and the country's population is stronger.

APPENDIX 9A

The fiscal strategies were surveyed in more or less the same way in the participating countries using the American questionnaire as reference. The exact wording of the 33 items (in the American questionnaire) was:

> Here is a list of fiscal management strategies that cities have used.... Please indicate the importance in dollars of each strategy (One of most important, Very important, Somewhat important, Least important, Don't know/not applicable).

In order to make the answers to the 33 items comparable across countries, we constructed the so-called 'Index of importance'. As a general rule, the index has been computed on the absolute frequencies as:

$$\frac{(\text{Very imp.} \times 3) + (\text{Somewhat imp.} \times 2) + (\text{Least imp.} \times 1) + (\text{Not used} \times 0)}{(\text{Very imp.} + \text{Somewhat imp.} + \text{Least imp.} + \text{Not used}) \times 3} \times 100$$

This procedure results in an index ranging from 100 (if all answered 'Very important') to 0 (if all answered 'Not used'). Note that 'Don't know' and 'Not applicable' and no answers have been left out of the computation.

In case four categories of importance were used, 'Most important × 4' has been added to the numerator, and 3 in the denominator has been replaced by 4.

Table 9.11 shows how the indices of importance were constructed for the various countries, including the weights assigned to the response categories. In some of the countries 'Not used' was not included in the survey, which means that 'Don't know/Not applicable/No answer' includes the strategies not used. This latter category includes strategies that could not have been used (not applicable) as well as strategies that could have been used but were not. When 'Not used' was not included

Table 9.11   *Index of importance for 33 austerity strategies*

	Most	Very	Somewhat	Less	Least	Not important	Not used	Don't know
Denmark	–	3	2	–	1	–	0	–
Norway	–	3	2	–	1	–	0	–
Sweden	–	2	1	–	–	–	0	–
Finland	3	2	1	–	0	–	–	–
France	–	4	3	2	–	1	0	–
UK	3	2	1	–	0	–	–	–
USA	4	3	2	–	1	–	0	–

as a response category, 'Least important' has been given a weight of 0.

APPENDIX 9B

The 10 figures that follow present a profile of each of the 10 countries covered in the book. Each diagram gives up to 16 pieces of fiscal information for municipal governments in each country, showing the actual number for the country as well as the average across the 10 countries. Most key figures depict trends over the period 1978–84 as percentage changes (in constant terms); a few depict the situation in a certain year. The 15 fiscal variables are defined as follows:

Fiscal slack: see Chapter 3 for precise definition.

Unemployment: rate of unemployment 1984.

Personal income: percentage change in personal income, 1978–84, fixed prices (deflator: consumer price index).

Tax base: percentage change in municipal tax base, 1978–84, fixed prices.

Tax effort: difference in percentages between total municipal tax effort 1978 and 1984. Total tax effort is defined as total taxes divided by tax base.

Reliance on tax revenues: total municipal taxes as a percentage of total revenues in 1984.

Fees and user charges: percentage change in municipal fees and charges, 1978–84, fixed prices.

Reliance on user fee: municipal fees and charges as a percentage of total revenues in 1984.

Total grants: percentage change in total grants to municipalities, 1978–84, fixed prices.

Current expenditure: percentage change in current municipal expenditures, 1978–84, fixed prices.

Capital expenditure: percentage change in capital municipal expenditures, 1978–84, fixed prices.

Capital/total expenditure: municipal capital expenditures as a percentage of total expenditures in 1984.

Municipal employees: full-time equivalent employees of the municipality.

Municipal wages: percentage change in total municipal wages, 1978–84, fixed prices.

Wage change: percentage change in 'municipal wages divided by number of employees', i.e. a rough measure of change in salaries per employee.

For the exact sources, cf. Mouritzen and Nielsen (1988: 26–52 and Table 2.1 of Country Index Tables for the respective country).

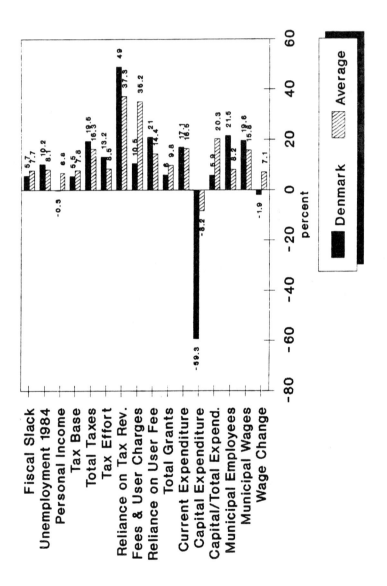

Figure 9.1   *Percentage changes, Denmark, 1978–84 (Mouritzen and Nielsen, 1988)*

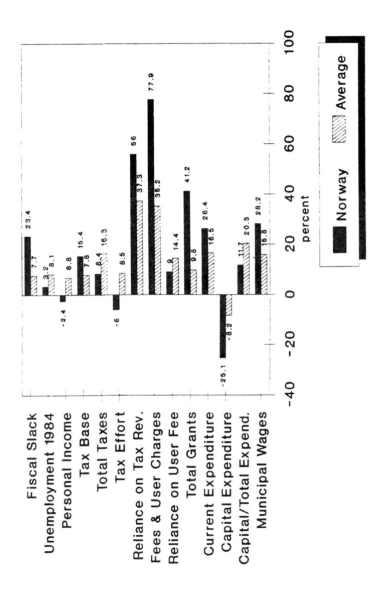

Figure 9.2 *Percentage changes, Norway, 1978–84 (Mouritzen and Nielsen, 1988)*

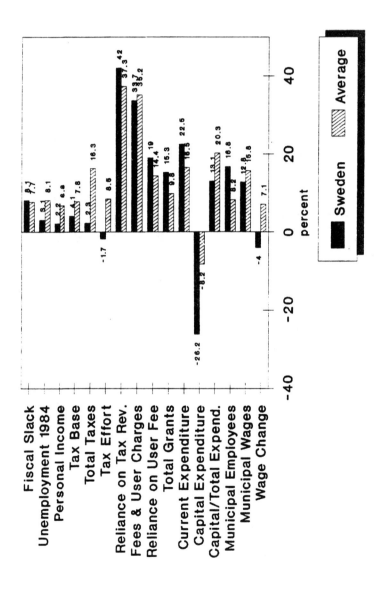

Figure 9.3   *Percentage changes, Sweden, 1978–84 (Mouritzen and Nielsen, 1988)*

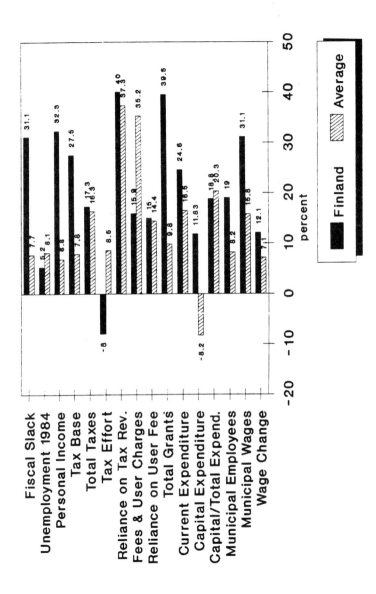

Figure 9.4 *Percentage changes, Finland, 1978–84 (Mouritzen and Nielsen, 1988)*

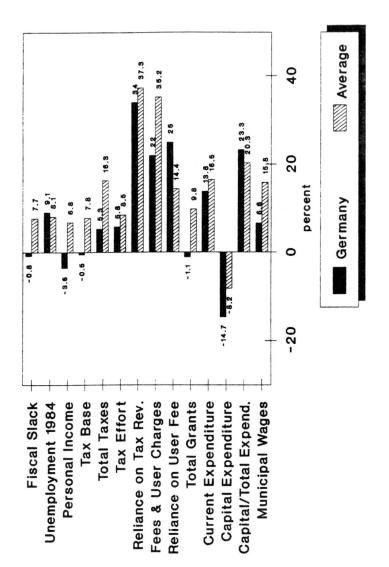

Figure 9.5   *Percentage changes, Germany, 1978–84 (Mouritzen and Nielsen, 1988)*

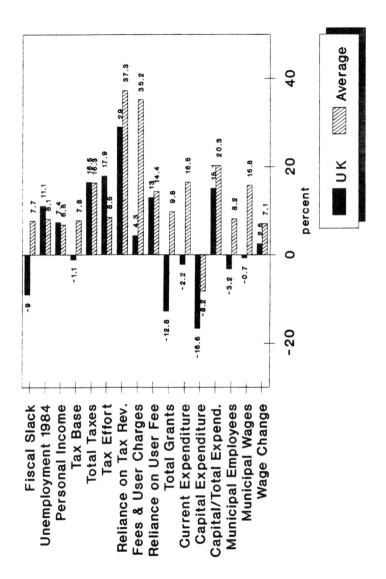

Figure 9.6 *Percentage changes, UK, 1978–84 (Mouritzen and Nielsen, 1988)*

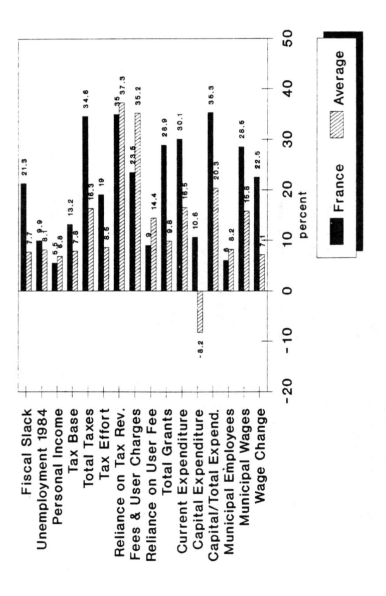

Figure 9.7   *Percentage changes, France, 1978–84 (Mouritzen and Nielsen, 1988)*

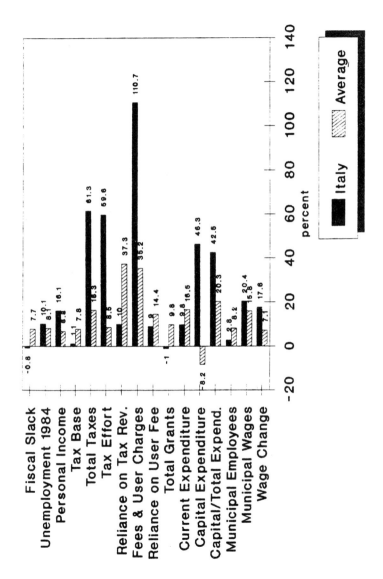

Figure 9.8   *Percentage changes, Italy, 1978–84 (Mouritzen and Nielsen, 1988)*

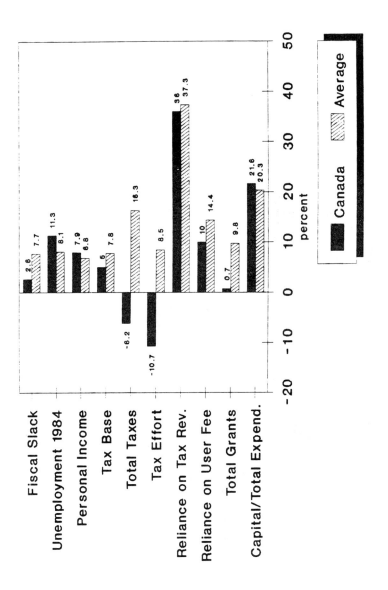

Figure 9.9   *Percentage changes, Canada, 1978–84 (Mouritzen and Nielsen, 1988)*

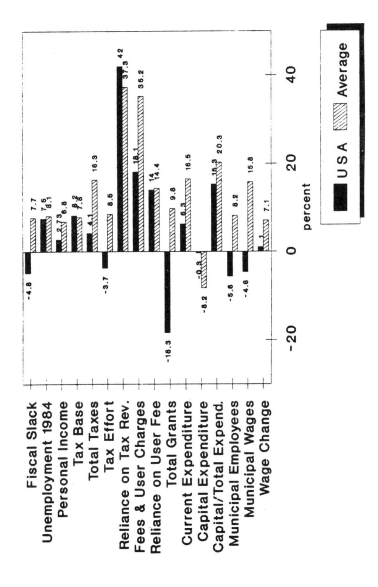

Figure 9.10 *Percentage changes, USA, 1978–84 (Mouritzen and Nielsen, 1988)*

REFERENCES

Appleton, Lynn M. 1989. 'Determinants of Innovation in Urban Fiscal Strategies', in *Research in Urban Policy*, vol. 3, Greenwich, Conn.: JAI Press, pp. 51–72.

Bahl, Roy 1969. *Metropolitan City Expenditures: A Comparative Analysis*, Lexington, Ky: University of Kentucky Press.

Baldersheim, Harald and Sissel Hovik 1987. *Does Leadership Matter?*, Bergen: Institute of Administration and Organization Theory.

Boyne, George A. 1985. 'Review Article: Theory, Methodology and Results in Political Science – The Case of Output Studies', *British Journal of Political Science*, 15: 473–515.

Brazer, Harvey E. 1959. *City Expenditures in the USA*, New York: National Bureau of Economic Research.

Burchell, Robert W. and David Listokin (eds) 1981. *Cities Under Stress: The Fiscal Crises of Urban America*, New Brunswick, NJ: Rutgers University Press.

Clark, Cal and Oliver B. Walter 1986. 'City Fiscal Strategies: Economic and Political Determinants', in *Research in Urban Policy*, vol. 2, Greenwich, Conn.: JAI Press, pp. 89–114.

Clark, Terry Nichols and Lorna Crowley Ferguson 1983. *City Money: Political Processes, Fiscal Strain, and Retrenchment*, New York: Columbia University Press.

Davis, Otto A., M.A.H. Dempster and Aaron Wildavsky 1966a. 'On the Process of Budgeting: An Empirical Study of Congressional Appropriation', *Public Choice* 1: 63–132.

Davis, Otto A., M.A.H. Dempster and Aaron Wildavsky 1966b. 'A Theory of the Budgetary Process', *American Political Science Review*, 60: 529–547.

Dye, Thomas R. 1966. *Politics, Economics and the Public: Policy Outcomes in the American States*, Chicago: Rand McNally.

Fabricant, Solomon 1952. *The Trend in Government Activity Since 1900*, New York: National Bureau of Economic Research.

Fried, R. 1975. 'Comparative Urban Policy and Performance', in F.L. Greenstein and N. Polsby (eds), *The Handbook of Political Science*, vol. 6, Reading, Mass: Addison-Wesley, pp. 305–379.

Levine, Charles H. 1980. *Managing Urban Stress: The Crisis in the Public Sector*, Chatham, NJ: Chatham House.

Mouritzen, P.E. and K.H. Nielsen 1988. *Handbook of Comparative Urban Fiscal Data*, Odense: Danish Data Archives, Odense University.

Niskanen, William A. 1971. *Bureaucracy and Representative Government*, Chicago: Aldine.

Sharpe, L.J. and Kenneth Newton 1984. *Does Politics Matter? The Determinants of Public Policy*, Oxford: Clarendon Press.

Skovsgaard, Carl-Johan 1985. 'Budgeting Innovations in Danish Municipalities under Fiscal Austerity', in Terry Clark, Gerd Michael Hellstern and Guido Martinotti (eds), *Urban Innovations as Response to Urban Fiscal Strain*, Berlin: Verlag und Versandbuchhandlung Europische Perspektiven GmbH, pp. 129–138.

Walzer, Norman and Warren Jones 1988. 'Spending Trends in American Cities', presentation to Midwest Political Science Association meeting, Chicago.

10

Municipal Employees and Personnel Policies: A Comparison of Seven Countries

Vincent Hoffmann-Martinot

Municipal employees come under continued scrutiny. City bureaucracies are often huge organizations and very important in policy decisions – New York City has 400,000 employees and Paris more than 40,000. Frequently, the municipal administration is the largest employer in the metropolis, or even in the region. Citizens perceive elected officials not only as service providers, but also as potential employers, able to intervene directly in the war against unemployment by hiring new municipal workers. The Fiscal Austerity and Urban Innovation (FAUI) project analyzed variations in decision-making processes and types of municipal personnel policies implemented in the 1980s. This chapter, by describing and comparing seven countries (Denmark, Norway, Sweden, Finland, the UK, France and the USA), answers three main questions:

1 Who are the municipal employees, and how did their numbers evolve during the last years?
2 What types of influence do municipal employees exert individually and collectively on local public policies?
3 Which personnel policies were adopted by municipalities in the 1980s, and under which conditions were they pursued?

GROWTH AND STABILIZATION OF THE MUNICIPAL WORKFORCE

Since 1945, the development of the welfare state has brought a rapid growth of national and state/local public administrations. The percentage of public employees in the civilian labor force increased rapidly in several countries between 1951 and 1981 (Table 10.1). Growth was particularly significant in Sweden (23 percent) and Denmark (22 percent), where today, as in France and the UK, about one-third of the active population works in the public sector (41 percent in Sweden in 1983, according

Table 10.1   *Public sector employees as a percentage of the total workforce*

	1951	1981	Increase, 1951–81
Denmark	9	31	22
Norway	–	–	–
Sweden	15	38	23
Finland	12	25	13
UK	27	31	4
France	18	33	15
USA	17	18	1

Source: Rose (1985); Bentzon (1987); Alestalo and Uusitalo (1988).

to Gustafsson, 1986: 42). In contrast, the relative importance of public employees has been stable in the USA at below 20 percent, reflecting a very different perception of the public sector's role in American society ('nightwatchman state').

Scandinavia   Among the seven countries, Scandinavian countries show the highest public employment. Following the reign of Gustav-Adolf and Sweden's entry into the Thirty Year War, the Swedish bureaucracy became one of the most competent and largest in the world (Peters, 1985a). The Swedish state has broad public responsibility for social activities and services, based on the principle that individual welfare is rooted in the collectivity. Such a global system covers not only targeted groups, but the entire population, and implies redistribution in a large number of sectors: housing, education, leisure and other services (Allardt, 1986; Esping-Andersen and Korpi, 1987).

Among Scandinavian nations, Sweden is closest to the institutional model of a welfare state. Not financed by contributions and covering all citizens belonging to a certain age category, its pension scheme is completely public, compared with a public–private mix in Denmark. Finland, too, differs from Sweden, with a more mixed socio-economic system (for instance in the housing sector) and a higher degree of political polarization: Social Democrats have to negotiate with Agrarians of the Center Party for the elaboration and implementation of major welfare state programs (Alapuro, 1982; Ylönen, 1985, 1986). Norway seems closest to Sweden in public sector development.

Most services linked to the growth of the welfare state in Scandinavian countries (education, health and social assistance) are supplied by local governments, and public employment growth was greater at the local than at the national level. Currently, one-half of the public sector employees

Table 10.2   *Local and regional employees as a percentage of the public sector*

	1951	1984	Increase, 1951–84
Denmark	–	57	–
Norway	–	64[1]	–
Sweden	29	54	25
Finland	–	–	–
UK	24	38	14
France	14	23[2]	9
USA	42	70[2]	28

[1] State and local employees only.
[2] 1980.

Source: Page and Goldsmith (1987), Rose (1985).

(including central administrations, public enterprises and local administrations) are in local administrations. (Cf. Table 10.2.) In Denmark, between 1972 and 1983 local government employment more than doubled (from 210,000 to 468,000), while central administration employment increased only one-tenth (from 180,000 to 200,000) (Tonboe, 1986). In Sweden, the 25 percent local employment increase in the public sector over this period was spectacular (Peters, 1985a).

Considering municipalities, rather than all local governments, full-time equivalents increased nearly one-fifth between 1978 and 1984: 22.9 percent in Denmark, 22.2 percent in Finland and 17.7 percent in Sweden (cf. Table 10.3).

The number of full-time equivalents (FTEs) per 1,000 inhabitants in the Nordic countries is unique (see Table 10.4). Employment per 1,000 residents increased about 10 percentage points during the period, while it stabilized or decreased in the non-Scandinavian countries. On the other hand, it reached approximately 60 FTEs per 1,000 Nordic residents in the mid-1980s, far above the USA or France (14), or even the UK (40). Data are not available for Norway, so Sweden leads the Scandinavian group with 64 FTEs per 1,000 residents in 1986.

Several factors explain the importance of municipal employment in the Nordic countries. Above all, the steady growth in the welfare state and range of functions assumed by consolidated municipalities is important. So are the amalgamation reforms implemented in the 1950s and 1960s, which, as in other European countries such as West Germany, strengthened municipal government bureaucratization.

UK   In Britain, also, there was strong growth in local government employment, much greater than in central government. In spite of a

Table 10.3 *Numbers of municipal employees'[1] and annual percentage increase*

	1978	1979	1980	1981	1982	1983	1984	1985	1986
Denmark									
Mun. empl.	216,167	233,000	249,506	260,427	266,615	268,132	265,565	265,697	269,765
% change	–	+7.8	+7.1	+4.4	+2.4	+0.6	–1.0	+0.1	+1.5
Norway[2]									
Mun. empl.	–	–	–	–	–	–	–	84,240	–
% change	–	–	–	–	–	–	–	–	–
Sweden									
Mun. empl.	441,042	462,306	488,368	497,623	494,702	507,404	519,003	524,472	533,700
% change	–	+4.8	+5.6	+1.2	–0.6	+2.6	+2.3	+1.1	+1.8
Finland									
Mun. empl.	246,319	254,879	261,131	270,586	287,117	293,599	301,121	–	–
% change	–	+3.5	+2.5	+3.6	+6.1	+2.3	+2.6	–	–
UK									
Mun. empl.	2,068,700	2,103,300	2,078,900	2,039,400	2,009,900	2,016,400	2,015,200	2,013,300	–
% change	–	+1.7	–1.2	–1.1	–1.5	+0.3	–0.1	–0.1	–
France									
Mun. empl.	–	–	–	–	–	–	761,319	769,154	772,978
% change	–	–	–	–	–	–	–	+1.0	+0.5
USA									
Mun. empl.	2,165,000	2,189,000	2,166,000	2,111,000	2,088,000	2,060,000	2,090,000	–	2,181,000
% change	–	+1.1	–1.1	–2.5	–1.1	–1.3	+1.5	–	+4.4[3]

[1] Full-time equivalents. [2] Excluding Oslo. [3] 1984.

Sources: Mouritzen and Nielsen (1988); US Department of Commerce (1988a).

Table 10.4 *Number of municipal employees per 1,000 inhabitants*

	1978	1979	1980	1981	1982	1983	1984	1985	1986
Denmark	48.1	51.6	55.0	57.3	58.7	59.0	58.5	58.5	59.2
Norway	–	–	–	–	–	–	–	–	–
Sweden	53.3	55.8	58.8	59.8	59.4	60.9	62.3	62.9	63.9
Finland	51.9	53.5	54.6	56.4	59.6	60.4	61.7	–	–
UK	41.8	42.5	41.9	41.1	40.5	40.6	40.5	40.3	–
France[1]	–	–	–	–	–	–	13.8	13.9	14.0
USA	15.5	15.6	15.4	15.0	14.8	14.3	14.3	–	–

[1] Except DOM–TOM (Départements d'Outre-Mer and Territoires d'Outre Mer).

Sources: Mouritzen and Nielsen (1988); US Department of Commerce (1988a).

transfer to the national administration of services traditionally delivered by local authorities (for instance, health services in 1974–75), the importance of local employment in the public sector grew from 24 to 38 percent between 1951 and 1984 (Table 10.2). Local public employment increased by 62 percent between 1961 and 1976 (from 1,870,000 to 3,022,000), with the major growth in education (from 785,000 to 1,569,000) (Thomson, 1982). In contrast to the Nordic countries and most other European countries, however, this steady increase did not continue after the mid-1970s. Central government austerity policies, increasingly applied by local governments, led to a stabilization and even a reduction in the number of jobs, especially in education, where 63,000 positions disappeared between 1979 and 1984. Between 1978 and 1985, local public employment was reduced by 2.7 percent (Table 10.3), and municipal employment per 1,000 residents diminished from 41.8 to 40.3 (Table 10.4).

France French public administration is much more centralized, with only 23 percent of public employees in local governments in 1980, while municipal employment per 1,000 inhabitants reached only 14.0 in 1986 (Table 10.4). Many functions assumed by municipalities in northern Europe are provided by *départements* and territorial states in France. These agencies are an intermediary level between central government and local governments; such governmental administration does not exist in the UK or the Nordic countries.

A first impression of strong and so-called 'typically French' centralization should be corrected. Indeed, when available data are examined closely, the 'jacobine France' has been decentralizing for many years. The main difference in comparison with the other countries is education. The French Education Ministry is often depicted as the largest European employer – after the Red Army: in 1987 this Ministry employed no fewer than 1,060,000 people, or 48 percent of the civil state employees (INSEE, 1988a).

If education in France were transferred to local authorities, as in many Western countries, the ratio would be radically changed. Local administrations would be employing 40 percent more people full-time than state agencies, instead of 44 percent fewer. In any case, the relative importance of local employees in the public sector increased by 9 percent between 1951 and 1980, and has, without doubt, increased since then (Table 10.2). In particular, a transfer from the state to local governments of new responsibilities and a central government policy to stabilize the number of state employees caused the latter to decline (in 1985 by 3,900 FTEs and in 1986 by 12,300 FTEs) for the first time since 1946. The rise in total municipal employment – which constitutes 80 percent of local employees – has been rapid since 1967. The number of municipal employees (excluding Paris and DOM–TOM) stood at 393,839 in 1967; by 1983 the figure had grown by 112 percent to 835,473. Since the mid-1980s, growth in the number of local employees (full-time equivalents) has slowed down. While the average annual increase was 4 percent between 1979 and 1983, it steadily diminished from 1983 onwards: 1.8 percent in 1984, 1.1 percent in 1985 and 0.4 percent in 1986.

USA   The level of public employment in the USA remains lower than in the European countries. Relative to the civilian labor force, public employment increased by only 1 percentage point between 1951 and 1981, from 17 to 18 percent (Table 10.1). Except during war periods, public employment has been concentrated much more in state and local governments than in the federal administration. This concentration is mainly attributable to education (employing one-third) and to the fact that federal programs include state and local jurisdictions: in most states, education employees (4,600,000 in 1986, according to the US Department of Commerce, 1988b) are employed by the school districts. The percentage of local employees in the public sector steadily increased between 1949 (50.3 percent) and 1977 (59.1 percent) (ACIR, 1986). Between 1970 and 1980, while federal employment remained stable (0.6 percent increase, from 2,881 to 2,898 million), state (up 36.2 percent, from 2,755 to 3,753 million) and local government employment (up 29.4 percent, from 7,392 to 9,562 million) grew rapidly. Municipal employment rose sharply. The workforce in New York City grew from 200,706 in 1961 to 294,522 in 1975, an increase of nearly 50 percent (Shefter, 1985: 117). As in Europe, a period of strong liberalism and local public intervention in the 1960s was followed from the mid-1970s onwards by a reduction in public sector growth, leading to a slower rise in local public employment increases. Declines of 1.9 percent in 1981 and

0.6 percent in 1982 (ACIR, 1986) were experienced, with a decrease in municipal employment of 5.9 percent (129,000 jobs) between 1979 and 1983 (Table 10.3; Lewis, 1988).

INFLUENCE OF MUNICIPAL EMPLOYEES ON LOCAL POLICIES

Following the distinction proposed by Banfield and Wilson (1967: 207), it is assumed that municipal employees occupy three main political roles: first, they are citizens with specific interests and preferences; second, they are employees organized in pressure groups; and finally, they participate directly in the development and implementation of municipal policies.

Specificity of interests and preferences of municipal employees

Blais and Dion (1987: 78) recently noted: 'it has to be admitted that the public employee generally is a badly known and in fact scarcely studied actor'. Nevertheless, several national studies have revealed the unique electoral behavior of public employees. In North America as in Europe, their participation in national or local elections is actually higher than that of private employees (on the USA: see Bennett and Orzechowski, 1983; for other countries, see Lipset, 1981: 191–192). Likewise, in most industrialized nations public employees more often favor left-wing candidates than do private employees. In the UK, Whiteley (1986) shows that the Labour vote is 10 percent higher among public than private employees. The difference was also clear in France in the second round of the 1988 presidential election, with 74 percent of the voting public employees choosing François Mitterrand, who was supported by only 60 percent of private sector voters (*Le Monde*, 'Dossiers et Documents: L'élection présidentielle', May 1988).

More precisely, and according to several studies, the higher the position in the professional hierarchy, the wider the orientation gap between public and private employees. The main finding of a 1978 French survey by the CEVIPOF/SOFRES is that 79 percent of private engineers and managers favored right-wing candidates, compared with only 42 percent of their counterparts in the public sector.

Based on results from electoral studies tracing a particular segment of public employees (strong participation and largely left-oriented vote), partisans of the public choice approach (Borcherding, 1977; Niskanen, 1971; Tullock, 1970) try to explain how bureaucrats 'naturally' behave as budget-maximizers. That analytical model has been criticized, especially on two points: its overestimation of the actual influence of bureaucrats in the decision system, and its ignorance of their attempts to rigorously manage the services for prestige reasons in the name of the 'general interest'. Rubin (1982) explains, for instance, how municipal employees in Southside, whose professional ethics emphasized quality rather than quantity, opposed a municipal council project to hire unqualified personnel. Moreover, the numerous empirical tests of public choice theory are not convincing, and are even contradictory. In Quebec, research of Blais and Dion (1987) does not indicate significant differences in expenditure preferences between the two groups, contrary to the results presented from Sweden by Sårlvik and Holmberg (1985) and from Denmark by Mouritzen (1987). American analyses of expenditure preferences expressed in local referenda regarding limitations on tax and expenditure increases show that, on average, public employees favor retrenchment measures less than private employees. They also show that their impact on the overall results is limited (Gramlich and Rubinfeld, 1982; Ladd and Wilson, 1983). Finally, the results of the recent Norwegian study by Hansen and Sorensen (1988) contradict the public choice proposition, because municipal administrators are less in favor of public intervention than elected officials.

Clientelism   One of the clearest expressions of the responsibility of elected officials in budget and personnel maximization is clientelistic politics. Evidently under various forms, this policy, through which a voter receives a job in exchange for his political support, is a common practice in many regions of the world (Roniger, 1981). Clientelistic hiring is a traditional characteristic of the French local political system. This phenomenon was described in detail by Jean-Yves Nevers (1983) in a study of the city of Toulouse under the Third Republic. During that period municipal jobs appear to constitute one of the fundamental resources of the clientelistic radical socialist system. Most hirings are concentrated at the lower end of the administrative hierarchy (sweepers, charwomen, etc.), not only within municipal services, but also in the various para- or intermunicipal services, or even outside the municipal sector (especially in the hospitals, whose executive councils are chaired by mayors).

Marseille is another example. In many respects, it resembles large American cities with a high degree of migration, an ethnic heterogeneity of population and integration and assimilation attempts. At the end of the last century, ethnic networks and clans (Corsican, Italian, Armenian) began to infiltrate all sectors of the Marseille social and political life ('clanism'). In the 1930s the Sabiani municipality was characterized by unbridled clientelism, hiring thousands of new employees between 1931 and 1935 (Bergès, 1984). When elected mayor in 1953, Gaston Defferre maintained his clientele, adopting the local saying, 'a service, a vote'. In the mid-1960s, a period of strong demographic rise, Marseille controlled approximately 50,000 jobs in municipal administration, transport authority, port enterprises, various public/private firms, social aid bodies, hospitals, housing agencies and other departments. In Nice, too, clientelistic networks – centering on the Medicine family – played an essential role, as they did in Draguignan during the 25-year-long domination of Mayor Soldani, 'the King of Var'.

Clientelistic involvement is a traditional and common practice also in Corsica, where mayors usually control permanent or temporary jobs. But, as often happens, patronage is controlled and organized mainly by political parties, particularly when such patronage is strongly structured. Within a municipal organization, the partisan logic may even compete with the objectives of the notable leader. Communist municipalities use clientelistic methods, but in a less systematic manner than often assumed; Dion (1986) established that the proportion of employees belonging to the majoritarian party is about the same in socialist as in communist municipalities in the Paris region.

Many studies have examined clientelistic government in American cities. (See Clark and Ferguson (1983) for the different components of the ethnic political culture, and Woody (1982) for policies carried on by new black mayors.) Traditionally, machines have rewarded workers by massive hirings in the municipal administration. In the last century, public employees in New York City depended completely on the machine. As in European countries, the growing creation and diffusion of rules regulating public employee careers were not sufficient to prevent clientelistic approaches. Mayor Lindsay used them to a large extent during his second term, hiring up to 28,000 temporary workers and shifting them regularly to other positions (Shefter, 1985). Likewise, in Boston and and in San Francisco, Mayors White and Alioto did not hesitate to apply overt and often questionable clientelistic methods of personnel management (Ferman, 1985).

While clientelism is a major feature in the North American and southern European countries, it is almost totally absent in the group that is termed for the purpose of this book the 'northern European' countries. In the Nordic countries, for instance, it is quite uncommon to find any systematic distribution of rewards to residents in the form of jobs in exchange for political allegiance. Except for the top positions in the bureaucracy, it is very unusual for elected officials to interfere in personnel matters. To do so for the benefit of a political follower or (to the extent that it happens, probably more often) for the benefit of a friend would be considered a clearcut case of corruption toward which legal steps are the appropriate answer.

Municipal employees as an organized pressure group
In most countries, unionization is substantially higher in the public than in the private sector. This results from the greater average size of public units, the usually better and closer relations between political parties and

Table 10.5   *Percentage of union members in the civilian labor force*

	1960	1965	1970	1975	1980	1981	1982	1985
Denmark	59.6	60.8	64.3	68.4	75.2	73.9	–	–
Norway	–	–	–	–	62.9[1]	–	–	–
Sweden	–	–	72.3	78.1	87.7	87.0	–	–
Finland	–	–	–	–	80.0[2]	–	–	–
UK	44.2	44.2	48.5	51.1	53.1	49.9	47.8	–
France	20.5[3]	20.8	23.1	22.9	19.2	19.0	–	15.0
USA	22.3	24.5	25.4	23.7	19.9	19.0	17.8	16.1[4]

[1] Data for 1979.
[2] Estimated.
[3] Data for 1962.
[4] Data for 1984.

Sources: Kassalow (1984); Lipset (1986); Mouriaux (1986); Troy (1986).

unions, and better opportunities to influence policies, especially through voting. But although union membership represents the main type of municipal employee organization, union practices differ markedly from country to country, as is indicated by the level and change in unionization rates after 1960. Table 10.5 lists three groups of countries by importance of their union concentration – very high (Scandinavian countries), high (UK), or low (USA and France).

Scandinavia   Most economically active Scandinavians belong to a union. Contrary to changes observed in many industrialized countries, the proportion of union members in the labor force has risen since the mid-1970s. Between 1975 and 1981, in Denmark the proportion increased from 68 to 74 percent, and in Sweden from 78 to 87 percent – without doubt a world record for unionization. Strongly linked to the Social Democratic Party, the three great Swedish federations (respectively, LO – Landsorganisationen, TCO – Tjänstemannens, and SACO–SR – Sveriges Akademikers Centralorganisationen–Statstjänstemans Riksforbund) cover not only employees of the private and public sectors (the military included), but also shopkeepers and the self-employed, unlike organizations in non-Scandinavian countries (Peters, 1985a, b).

The power in a federation such as LO is reflected by the 'empire' it controls: its co-operative, the greatest building firm in the country, includes a huge printing-house and a leisure and tourism organization. Also, it controls 25 percent of the daily printed newspapers. In the private as well as the public sector, union representatives have extended rights in Scandinavian countries. In municipalities many facilities to promote their activities are at their disposal. Scandinavian union representatives are currently involved in personnel policy changes at the national and local levels. These include the national corporatist system of wage settlements, general work conditions, and pension schemes of local employees' associates in Denmark (Mouritzen, 1989; Tonboe, 1986), Sweden (Anton, 1974) and Finland (Kiviniemi, 1988). Still, not all decisions are made during these national negotiations. Agreements – even confrontations – also occur in every municipality. In Norway, unions generally participate in many municipal commissions whose 'team spirit' is well known (Eckstein, 1966: 140). In addition to conventional and traditional relations between employers and employees, union action includes strikes, even though these seldom occur. For instance, in May 1986 the Swedish SACO–SR federation appealed to thousands of local employees to strike, when threatened by a lock-out.

UK   In the UK, nearly one-half of the civilian labor force was unionized in the early 1980s. This high rate was partly the result of a closed-shop practice covering 5 million workers or half of all union members, and of the deduction in advance of membership subscriptions from wages. In the public sector, unions were created by the end of the nineteenth century: the Municipal Employees' Union (MEU) in 1894, the National Union of Teachers (NUT) in 1870 and the National and Local Government Officers' Association (NALGO) in 1905. After the 1960s and the 1970s, the strongest increase in union membership was in the public sector, including 12 of the 25 largest national unions in 1981 (Parry, 1985: 80). In the early 1980s, an estimated 80 percent of national or local public employees belonged to a union (Rose, 1985; Thomson, 1982).

Since 1945, wages and work condition settlements for local employees are negotiated nationally in the UK, but not in as consensual a manner as in the Nordic countries. The Local Authorities' Conditions of Service Advisory Board (LACSAB), a committee composed of representatives of ministries as well as local authorities and unions, regularly produces recommendations for employers. Analyzing in-depth the functions and operations of firefighters, Rhodes (1984, 1986) demonstrated how this national regulation system was essentially dominated by central government orientations. In fact, representatives of employers and employees – most importantly, after the electoral victory of the Conservatives in 1979 – played a crucial role in determining the pay and work conditions of local employees. The flexibility of unions is more important within local authorities, which control many aspects of personnel management (hiring practices, career opportunities of employees, determination of fringe benefits and so on).

Local unions have been increasingly active in the UK since the 1960s. First, there were strikes of the NALGO members in 1964, then a rapid extension of teacher demonstrations, ambulance workers and firefighters

(Hampton, 1987: 140). General principles adopted at national tripartite meetings are more often modified and adapted at the municipal level, specifying their content and considering local peculiarities; unions such as NALGO or NUPE (National Union of Public Employees) begin to designate shop stewards more frequently, and to claim the right to participate in local government commissions. Recently they have obtained a right to be consulted in some local authorities like Liverpool, Slough, Hereford and Worcester, and Basildon.

By the mid-1970s, union activism had developed in reaction to the adoption of austerity policies by the central government. Enlarging their involvement to more general political and economic issues, unions vigorously opposed measures reducing the growth in expenditures, local taxes and a transformation of local institutions (Gyford, 1985; Gyford and James, 1983).

Since 1980, a decline in the UK national unionization rate has been observed (Table 10.5). Has the rate fallen in local authorities, also? Mobilizations and actions implemented for years have produced only limited effects (Pickvance, 1986). The ability of unions to influence policies was gradually reduced by the Thatcher government. The 1980 Employment Act limited the practice of strike pickets and solidarity strikes. The 1982 Employment Act strictly defined legal strikes and reduced union immunity in conflict cases. The 1988 Employment Act declared that a strike can be organized by a union only after a secret ballot. Since 1979, attacks by the government against 'corporations' have led to a weakening in the traditionally strong influence of professionals (especially in the housing sector), who are nationally organized but benefit local authorities with their competence and expertise (Goldsmith, 1988; Pickvance, 1985a).

USA   The USA is in a third group of countries, characterized by a low national level of unionization (between 15 and 30 percent). However, without the rapid increase in public unionism since the 1960s, the rate of 16 percent in 1984 would have been lower. Indeed, between 1953 and 1976, while the percentage of union members in the private sector declined (from 36 to 25 percent), in the public realm it increased (from 12 to 40 percent) and mainly at state and local levels (cf. Table 10.6).

In the 1960s and early 1970s the difference between the two sectors lessened, not only in size and activism, but also in rights attributed to unions. Executive Order 10988, created in 1962 by President Kennedy, produced an impact in the public sector as important as that induced in the private sector by the Wagner Act. In the mid-1970s, 40 percent of municipal employees were organized – either in unions (20 percent) or in professional associations (20 percent) – and nearly all of them in large cities (Lineberry and Sharkansky, 1978). More and more, the employees were affiliated with AFL–CIO or with the American Federation of State, County and Municipal Employees (AFSCME), the International Association of Firefighters (IAFF) or the American Federation of Teachers (AFT). Also, in a more traditional way, they are affiliated with professional associations such as the Fraternal Order of Police, the Uniformed Firemen's Association and the National Educational Association, or even with confessional fraternities such as the Holy Name Societies, the St George Societies and the Shomrim Societies (Sayre and Kaufman, 1965: 75). Contrary to European countries such as the UK, the organization of American municipal employees as an independent force is rather new, having developed only in the past 20 years.

Another uniqueness of the American case is the primarily local decision-making regarding personnel, including wages. For municipal employee unions, the city hall is the sole action and bargaining arena, although unions try to influence state legislative processes that create the work rules in the local public sector. If the political weight of unions in policy processes has, on the whole, increased in American cities since the 1960s, it varies widely from one municipality to another according to various factors, including the centralization of political leadership. According to Banfield and Wilson (1967), the more the party and the executive are centralized, the more likely they are to resist pressures of organized employees or the dominant political culture ('pro-labor' in Boston, 'anti-labor' in San Diego) (Clark et al., 1984). In many cities there were strikes in the 1960s, with the number of striking employees increasing from 28,000 to 252,000 between 1960 and 1975 (Clark and Ferguson, 1983: 154). For example, in Chicago, after selection of a new union leader, firefighters spent as much energy fighting the successive mayors as they did fighting fires (Grimshaw, 1982). To prevent trouble

Table 10.6 *Private and public unionization in the USA*

	Employees ('000s)		Density[1] (%)	
	Private	Public	Private	Public
1953	15,540.2	769.8	35.7	11.6
1962	14,731.2	2,161.9	31.6	24.3
1970	16,978.3	4,012.0	29.1	32.0
1973	16,803.5	5,077.8	26.6	37.0
1976	16,166.8	5,980.3	25.1	40.2
1983	13,142.6	5,410.7	17.8	34.4

[1] Proportion of employees unionized.

Source: Troy (1986).

and a decline in their popularity among citizens, many mayors confronted with strike threats preferred to give up and accept employees' demands. So, while the fiscal situation in New York City worsened, Lindsay, as well as Beame, could not decide to adopt more stringent personnel policies (Shefter, 1985: 117–118).

Yet the power of unions in the USA has substantially decreased following the 1970s. The conservative wave, exemplified by movements toward expenditure and tax reduction, led to a change of mayoral attitudes regarding demands and strategies of municipal employees. Now they often resisted union demands, sometimes during difficult labor conflicts, as in Seattle, Atlanta or San Antonio. In San Francisco, while unions had been closely aligned with the Alioto administration, their importance was significantly reduced by the Moscone and Feinstein administrations (Ferman, 1985). In numerous cases, faced with acute fiscal problems, municipalities reacted with a reinforced centralization of politico-administrative leadership, which facilitated resistance to pressures by municipal employees (Levine et al., 1981). Compared with the 1960–75 period, unionization experienced a decline in numbers and militantism, marked by the decline of union importance in the public sector from 40 to 34 percent between 1976 and 1983 (Table 10.6) and the diminution of strikes (from 252,000 striking local employees in 1975 to 206,000 four years later) (Troy, 1986). Underlying this decline is the less pro-union orientation of many political leaders – more and more supported by residents. Opinion polls taken in the late 1950s and early 1980s reveal a growing disaffection of the American population with unions (Lipset, 1986; Lipset and Schneider, 1983). Also, municipal hiring included an increasing number of part-time employees, only a minority of whom favored unionization – 8 percent in 1982, compared with 46 percent of full-time employees (Lewin, 1986; Stein et al., 1986; Troy, 1986).

France   France is the least unionized country in our seven-country sample. Estimating the number of unionized employees precisely is difficult. According to Mouriaux (1983: 65), 'estimations given by unions are suspicious. If there is no doubt that in the past CGT and CFTC inflated their official statistics to improve their representativity, current data in any case remain doubtful.' Estimates in 1988 reached 23 percent. So today France, like Greece, is a European country where unionism is at a low.

Two other elements reflect a decline of the unions. For ten years, elections to the *comités d'entreprise* (negotiating bodies in private firms and public enterprises) have been marked by a rapid increase in the number of non-unionized candidates (24 percent of votes in 1987, an increase of 5 percent since 1977), with a steady decline of the largest union, the Confédération Générale du Travail (CGT) (27 percent of votes, 10 percent less), and a stabilization of the other unions (Confédération Française Démocratique du Travail – CFDT, Force Ouvrière – FO, Confédération Générale des Cadres – CGC and Confédération Française des Travailleurs Chrétiens – CFTC). Moreover, activism, as measured by the intensity of collective work conflicts, has diminished since the mid-1970s: one has to go back to 1946 to find a year with fewer conflicts than 1985.

The decline of union influence in French society may continue in the years ahead as, like the Americans, French citizens become more and more distrustful of unions – 57 percent of respondents to a SOFRES poll in 1985 said they had no trust in unions, compared with 51 percent three years before (SOFRES, 1987: 161).

This relative weakness in unionism cannot be explained by a 'national French uniqueness', according to which the French would be 'naturally' and basically independent and hostile to collective organization. Other factors determine the French characteristics also. Apart from the privileged relationship tying the Fédération de l'Education Nationale to the Socialist Party, links between unions and parties have never been as strong or symbiotic in France as in the northern European countries. As Wilson (1983: 247) describes it, 'a solid tradition isolates unions from parties'. In addition, closed-shop practices are illegal and survive in only two sectors, Paris printing and dockers. Finally, the potential influence of unions in France is strongly limited by a high degree of fragmentation. This division in particular can be found in the public sector, which resembles a mosaic of little corporative groups (Crozier, 1967). However, the membership rate (35 percent) remains higher there than in private firms. With greater involvement in labor negotiations in the public sector than in the private sector, unions such as the Syndicat National des Instituteurs in primary education or the Fédération CGT de l'Energie en Electricité de France have more control over public careers.

Vis-a-vis municipal employees, French unions, as early as the beginning of this century (from the 1910s to the 1940s), pursued a fruitful strategy of national negotiation, with associations of mayors and the state playing the arbiter's role. During that period, legislation was passed establishing national norms for every municipality for the hiring, promotion, wages and work conditions of employees (Lapassat, 1977; Thoenig and Dupuy, 1980; Thoenig, 1982).

In the realm of general economic policy as defined by the central government, wage scales are negotiated at the national level for the entire public sector, and local employee salaries are calculated in reference to those of state employees. At the national level, also, the influence of unions is more pronounced in the training sector, a central issue of corporatist negotiations between representatives of state and national associations of local elected officials (Thoenig and Dupuy, 1980). Therefore, certain personnel decisions are not under municipal control. Despite strong jurisdictional constraints, the ability of municipal officials to negotiate is not negligible, especially for policies relating to staff size, careers and fringe benefits. Union stewards participate in negotiation practices in the municipal administration including *commissions administratives paritaires* (career and disciplinary matters), *comités techniques paritaires* (services organization and functioning) and *comités d'hygiène et de sécurité* (health and security). As in the national public administration, action by unions is most often characterized by a structural division, which partly explains the frequent failure of fights against retrenchment and privatization attempts in many municipalities. One of the factors favoring the rapid implementation of Jacques Chirac's privatization program in Paris was the fragmented and therefore weak defence of employee interests in ten different unions.

Municipal employees as policy-makers
In the Weberian model of bureaucracy, political and administrative functions are strictly separated. The administrative employee loyally executes decisions of the political authority and 'has to carry out his work *sine ira et studio*, without resentment and bias' (Weber, 1963: 128). The American reform movement of the last century proposed isolating 'government' from 'politics', in order to prevent any risk of corruption or clientelism. But there is no doubt that a strictly defined separation of functions never matched reality, and numerous municipal employees act on the borderline of political and administrative spheres (Banner, 1982; also Jobert and Sellier, 1977).

Far from restricting their activity to pure administrative tasks, administrative officials often serve as major decision-makers. By their presence, expertise and professionalism, they directly intervene in various stages of the decision and implementation process. Their independence and autonomy are particularly important in the UK, where councillors' influence is sharply limited by methods used by administrators to monopolize information and relations with ministries (Byrne, 1986; Glassberg, 1980). In Norway, the 'Raadmenn' are also strong executives. In Denmark Mouritzen (1990) showed that spending and taxation tended to increase at a faster rate in municipalities characterized by strong bureaucrats and weak political leaders. At the top of the French municipal hierarchy, the general secretary frequently concentrates such power that he can reign over 'his' administration; in many municipalities he is significantly called 'le général'. Similarly, in the USA many studies have shown how the mayoral leadership may be challenged or limited by chief administrators, who oppose the mayor's instructions (in New York: see Banfield and Wilson, 1967: 218) or resist the introduction of new policies (in San Francisco: see Ferman, 1985).

However, since the 1970s the stronger involvement of many elected officials in municipal management has led, in many countries, to a redefinition of the spheres of politics and administration and to a balance conforming to the normative Weberian model. 'Dilettantes' have come so much closer to 'specialists' that they compete or even substitute. That specialization or professionalization, 'meaning to have a technical competence, an expertise' (Mabileau and Sadran, 1982: 269) is observed through the development of full-time activities among half of the mayors in Norway (Larsen, 1987), as well as in French municipalities with more than 2,000 inhabitants, according to the results of a survey conducted in June 1988 by CSA/Le Monde. The prospect of making a city administration more transparent, accessible and visible induces elected officials to be more active in the *départements* and so to prove through their various and daily interventions that they favor a more dynamic policy style in economic crises.

The relationship between a crisis environment and dynamic leadership can be found in Western countries following the mid-1970s (Goldsmith, 1987: 5). The new rules of the game necessarily produce a reactivation of the 'fundamental tension' (Mayntz, 1982: 63) characterizing the relationship between 'dilettantes' and 'specialists'. Elected officials know they can introduce change only with the support of, or at least non-resistance by, employees. They have at their disposal many resources to oppose plans of political leaders by using, manipulating or curtailing internal information through various decision-making techniques (Sayre and Kaufman, 1965: 421), or modifying and limiting the implementation. Unco-ordinated interventions by elected officials in 'core questions' of administration can lead to such serious withdrawal and blocking-up reactions of a majority of administrators that conditions of the pre-bureaucratic administration described by Weber are likely to re-emerge. Even today, numerous employees have limited qualifications, are involved in routine projects and fit the Downesian model of conservative public employees (Downs, 1967). Their behavior is characterized by a penchant for security and opposition to change. Nevertheless, a growing number of better qualified, younger and more competent people have been hired in recent years and contribute to a renewal of bureaucratic methods and practices. Even if they face the inertia of conservative colleagues, who are basically hostile to the modification of existing structures and to the competition induced by this 'new blood', those change advocates (using Downs' terminology) will be active agents of the administration's modernization.

MUNICIPAL PERSONNEL STRATEGIES AND POLICIES

This final section presents the survey results from the Fiscal Austerity and Urban Innovation (FAUI) project concerning strategies directly affecting personnel and strategies relating to staff, compensation and privatization. Data for the seven countries are translated to the so-called 'index of importance' (Mouritzen and Nielsen, 1988; cf. also Appendix 9A above), from chief administrative officer responses. The index can take values from 0 to 100. A value of 0 indicates that no municipality in the country reported having used the strategy; a value of 100 indicates that all responding municipalities reported having used the strategy and that it had been very important. The index thus reflects both the frequency of usage of these strategies and their fiscal impacts.

Strategies relating to staff

Results for the four strategies aimed at reducing or stabilizing staff – layoff, hiring freeze, workforce reduction through attrition and retirements – show two distinctive types of policy among the Nordic countries (Table 10.7). In Norway and Sweden, layoffs and early retirement are seldom used, nor are other strategies (except for reducing the workforce through attrition in Swedish cities). The high degree of fiscal slack, the dominant ideology of the welfare state and strength of unions are the main factors impeding the adoption of staff reduction strategies. By comparison, Denmark and Finland are closer to non-Scandinavian countries, especially in their relatively high level of layoffs (26 and 30). In the Danish case this is probably due to the timing of the survey, which was carried out in the spring of 1983, a few months after the first and most extensive round of personnel cutbacks in the 1980s (forced upon local governments because of reductions in grants after the passing of the 1983 budget; cf. Chapter 5). Finland is distinct from the 'Scandinavian model' exemplified by Sweden or Norway, because of a greater polarization of its partisan system – preventing social democracy from playing a dominant

Table 10.7   *Strategies concerning the numbers of municipal employees*

	Lay off personnel	Freeze hiring	Reduce workforce through attrition	Introduce early retirements
Denmark	26	51	–	–
Norway	4	22	15	2
Sweden	0	5	43	3
Finland	30	40	–	22
UK	26	37	21	36
France	4	70	41	–
USA	30	41	49	13

Source: Mouritzen and Nielsen (1988).

role – which is reflected within the unions, weakened by divisions pitting communists against social democrats.

In France, strategies limiting growth in the number of local employees were implemented after the 1983 municipal elections. One of the most striking cases of the shifting direction of personnel policies is found in Paris. While during his first term Jacques Chirac very much favored strong public interventionism and an extension of municipal staff, hiring hundreds of new employees, he then adopted retrenchment practices. Nevertheless, layoffs in France were few (Table 10.7). Most municipal employees have a long tenure, and the rare layoffs are mainly for contractual services, often politically involved with previous local leaders. Municipalities commonly stabilize bureaucracies with a hiring freeze, and therefore the value of the importance index for this strategy is highest in France. Fewer tenure jobs are created, compared with the development of temporary and auxiliary positions (especially under the form of the Travaux d'Utilité Collective – TUC). But personnel policies vary widely and do not necessarily fit in with official political ideologies. Though a conservative leader, the mayor of Fréjus, François Léotard, created no fewer than 13 divisions and doubled the number of employees (from 330 in 1977 to 610 in 1986) following his election to city hall. Inversely, many left-wing municipalities pursue very restrictive policies (see the example of the communist municipality of Tarbes in Clark et al., 1987: 361–390). Between 1983 and 1986, Pierre Mauroy, the Socialist mayor of Lille and former prime minister, against strong reactions of unions, reduced the number of municipal employees by 10 percent.

Since the mid-1970s, the British central government has tried to reduce employment in local authorities. Under this pressure, many authorities did reduce employment, but without using massive layoffs. (In Wandsworth a reduction of roughly 700 jobs – 9 percent of the workforce – took place in 1981, but this example remains relatively unique: Chandler and Lawless, 1985: 160; Wolman, 1983). Alternative policies, such as reductions by attrition in education, often negotiated with unions, were implemented instead. The combined effects of such measures led to a rapid decrease in the number of teachers – 25 percent – in Torytown (Duke and Edgell, 1986). Compared with other countries, early retirement is a unique strategy in the UK (Table 10.7). Concerned about growing unemployment, many local authorities intervened actively in the labour market, not by following the politically dangerous example of the radically oriented council of Liverpool, but through job-sharing, a means of dividing a full-time position into several part-time jobs and reducing the influence of unions: part-time employees more frequently are women and non-union members (Karran, 1984).

In the USA, many case studies and statistical analyses underline the link between intergovernmental transfers and municipal workforce size

(Schneider, 1988). Numerous local governments used the Comprehensive Employment and Training Act (CETA), passed in 1973, to finance new jobs (Woody, 1982), or even the rehiring of illegally laid off employees (Rubin, 1982: 105). When municipal officials thought that federal and state aid had become a regular and secure flow to finance job creation, transfers began to be reduced (Stein, 1984). Diminution of intergovernmental grants, fiscal crisis and growing conservatism induced a fall in local employment. In cities such as Detroit or New York, faced with acute fiscal problems, spectacular layoffs occurred. Ethnic minorities, numbers of whom had been hired since the 1960s, were major victims of these austerity policies, according to the rule of 'last hired, first fired'. In New York, the number of white employees declined by only 22 percent between 1974 and the end of 1975, but the numbers of black and hispanic employees declined by 35 and 50 percent, respectively (Shefter, 1985; Eisinger, 1982b; Levine, 1980). Strongly organized groups resisted layoff measures more effectively. Certainly that is true of firefighters and policemen, who are among the most organized employees in the local public sector (71 and 53 percent in 1980, according to Gross, 1985: 411; see also Stieber, 1973; Stern, 1984). In austere as well as in prosperous times, firefighters and police officers most often succeed in negotiating skillfully the improvement or the non-reduction of their specific benefits (as illustrated by their repeated successes in Oakland: see Levine et al., 1981: 23). The radical practice of massive layoffs was replaced at the end of the 1970s by attrition, today reported by US cities as the most important personnel reduction strategy (Table 10.7). As in New York, many municipalities negotiated agreements with unions excluding layoffs but providing a personnel decrease, combined with a productivity increase (Lewin, 1986; Shefter, 1985).

In these seven countries, the stabilization or reduction of workforces is sometimes partially compensated by the use of volunteers. Following the fiscal crisis of 1975, New York City created a citizens' committee to train volunteers in various sectors: security, assistance to families, the elderly, migrants and housing renewal. The use of volunteers is much less common in France, where they add the equivalent of only 1 percent to the workforce, compared with 4 percent in the USA and 3 percent in the UK (Le Net and Werquin, 1985). One can argue that, the higher the trust between individuals, the more informal activities – help and solidarity – will develop. This explains why the use of volunteers is more limited in France – where inter-individual trust is particularly weak – than in Scandinavian countries, the UK and the USA.

Strategies relating to compensation

A second group of personnel strategies involves municipal employee compensation (Table 10.8). In Europe (with the possible exception of the

Table 10.8   *Strategies concerning the pay of municipal employees (index of importance)*

	Reduce employee compensation	Freeze wages and salaries	Reduce overtime
Denmark	–	59	–
Norway	11	41	14
Sweden	–	11	8
Finland	22	59	15
UK	13	8	32
France	–	31	35
USA	11	22	33

Source: Mouritzen and Nielsen (1988).

Swiss), individual municipalities have few options in wage settlements, because pay conditions are determined at the national level; nevertheless, they set work conditions and benefits. Again, Danish and Finnish cities differ from Norwegian and Swedish municipalities in using more restrictive measures, especially wage and salary freezes. British local officials favor reducing overtime. That strategy is almost as important in France, where many municipalities after the beginning of the 1980s tried to control increases in employee benefits.

To reduce absenteeism, the mayor of Lille opposed unions with a hiring freeze and, in 1984, restricted bonuses. In the same year, the communist mayor of Le Mans and his socialist deputy, president of the Communaute Urbaine du Mans, were sequestered during several hours for having limited employee benefits.

Contrary to the European cases, municipalities in the USA are much freer to set compensation levels. Yet Table 10.8 shows that the importance of compensation reduction strategies is most often weaker in American cities. Tables 10.7 and 10.8 indicate that cities confronted with fiscal stress preferred to use staff (rather than compensation) reductions. This result confirms several previous analyses (Clark and Ferguson, 1983; Lewis, 1988; Schneider, 1988; Stein, 1984; Wolman, 1983), which have shown that, supported by unions, most municipalities prefer to reduce the workforce rather than displease all employees by worsening their conditions.

Privatization of municipal services
Privatization of municipal services is a strategy with considerable effects on personnel. Privatization appears marginal in Norway and Sweden, where the political culture remains widely acceptant of the welfare state (Allardt, 1984, 1986; Flora, 1988). This differs from the less homogeneous politico-economic system in Finland, characterized by

stronger competition between private forces and the public sector (Table 10.9). Under central government pressure, privatization was used in the UK after 1979 by numerous (mainly Conservative) local authorities and often provoked serious strikes (Duke and Edgell, 1986; Walker, 1983).

In France, the privatization of municipal services was frequently advanced in 1983 by newly elected right-wing mayors as an 'antidote' to the socialization of the economy and society. Local experimentation was recommended before its adoption at the national level. Many right-wing (in particular, RPR) elected officials were suddenly converted to market principles and free enterprise and viewed privatization as a true panacea. But in some cities, severe privatization measures were adopted, and were applied with haste and with insufficient control. In other cities such as Nimes (Hoffmann-Martinot, 1988) unions opposed, sometimes violently, the 'dismantling of the public service'; but in most cases the privatization measures finally passed, with elected officials' agreements determining in detail the methods of shifting to the private sector.

Half of all American cities, mainly those under Republican control, use some privatized services. Numerous reports and studies convinced municipal officials of the low competitiveness of the public sector. Consequently, between 1975 and 1983 the number of employees (full-time equivalents) per 1,000 inhabitants in those municipal departments fell by 50 percent, from 1.4 to 0.7, according to Lewin (1986). Sanitation is one service in which privatization has increased rapidly. Lewis (1988) indicates that 5 percent of a sample of 154 large cities eliminated their sanitation department between 1979 and 1983 (Fitzgerald et al., 1988).

Comparing public and private costs is only one criterion that can be used by a municipality before changing service delivery mode. In many cases, resistance by organized employees has an equal or even greater influence in the final choice. In this respect, Ferris (1986) states that, the more organized are the personnel, the less likely is the implementation of privatization policies.

Table 10.9   *Contracting out of municipal services*

	Index of importance
Denmark	–
Norway	9
Sweden	6
Finland	39
UK	30
France	22
USA	27

Source: Mouritzen and Nielsen (1988).

CONCLUSION

This chapter is a sketch of a more systematic work underway to compare, across countries, the role of municipal employees in local policies. A brief review, in different national contexts, of the history and development of municipal bureaucracies, their evolution, relationships between administration and politics, types of organization and defense of local employees' interests was presented. We explained how and why certain personnel strategies were applied in some countries, but only marginally in others. From the previous analyses, some interesting results can be underlined.

1 In all countries, multiple strategies to stabilize personnel expenditures have been introduced.
2 The more numerous, organized and influential the municipal employees are, the more likely it is that they will manage to alleviate the impact of those strategies.
3 The degree of fiscal slack (cf. Chapters 3 and 4) has little effect on the selection of personnel policies. For instance, retrenchment policies adopted since the beginning of the 1980s in France result more from pressures exerted by state authorities and ideological orientations.
4 International comparisons can easily lead to biased results if homogeneous national types of 'regional models' are presupposed and serve as constant references. Intra-country differences must always be firmly kept in mind, in addition to historical and political traditions (significant differences within the block of Nordic countries), with regard to union implantation or urban political cultures (in the USA).

REFERENCES

ACIR 1986. *Significant Features of Fiscal Federalism*, 1985–86 edn, Washington, DC: ACIR.
Alapuro, Risto 1982. 'Finland: An Interface Periphery', in Stein Rokkan and Derek W. Urwin (eds), *The Politics of Territorial Identity: Studies in European Regionalism*, London: Sage, pp. 113–164.
Alestalo, Matti and Hannu Uusitalo 1988. 'Finland', in Peter Flora (ed.), *Growth to Limits: The Western European Welfare States Since World War II*, vol. 1: *Sweden, Norway, Finland, Denmark*, New York: Walter de Gruyter, pp. 197–292.
Allardt, Erik 1984. 'Representative Government in a Bureaucratic Age', *Daedalus*, 113(1): 169–197.
Allardt, Erik 1986. 'The Civic Conception of the Welfare State in Scandinavia', in Richard Rose and Rei Shiratori (eds), *The Welfare State East and West*, Oxford University Press, pp. 107–125.

Anton, Thomas J. 1974. 'The Pursuit of Efficiency: Values and Structure in the Changing Politics of Swedish Municipalities', in Terry N. Clark (ed.), *Comparative Community Politics*, New York: John Wiley, pp. 87–110.

Baldersheim, Harald 1988. 'Does Money Matter? Impacts of Financial Change on Local Politics in Norway', International Workshop on Local Finances in the Contemporary State: Theory and Practice, University of Oslo, 5–7 May.

Banfield, Edward C. and James Q. Wilson 1967. *City Politics*, Cambridge, Mass.: Harvard University Press.

Banner, Gerhard 1982. 'Zur politisch-administrativen Steuerung in der Kommune', *Archiv für Kommunalwissenschaften*, I: 26–47.

Bennett, James T. and William O. Orzechowski 1983. 'The Voting Behavior of Bureaucrats: Some Empirical Evidences', *Public Choice*, 41(2): 271–283.

Bentzon, Karl-Henrik 1987. 'A Very Short Note on Expenditures and Productivity in Danish Somatic Hospitals from 1950 to 1985', International Institute of Administrative Science, Conference on 'Public Administration in Times of Scarce Resources', Valencia.

Bergès, Michel 1984. 'Peut-on sortir de la corruption?' *Pouvoirs*, no. 31: 65–75.

Blais, André and Stephane Dion 1987. 'Les employés du secteur public sont ils différents?', *Revue Française de Science Politique*, 37(1): 76–97.

Bogason, Peter 1987. 'Denmark', in Edward C. Page and Michael J. Goldsmith (eds), *Central and Local Government Relations: A Comparative Analysis of West European Unitary Systems*, London: Sage, pp. 46–67.

Borcherding, Thomas E. 1977. *Budgets and Bureaucrats: The Source of Government Growth*, Durham, NC: Duke University Press.

Brunn, Finn and Carl-Johan Skovsgaard 1981. 'Local Determinants and Central Control of Municipal Finance: The Affluent Local Authorities of Denmark', in L.J. Sharpe (ed.), *The Local Fiscal Crisis in Western Europe: Myths and Realities*, London: Sage, pp. 29–61.

Byrne, Tony 1986. *Local Government in Britain: Everyone's Guide to How It All Works*, 4th edn., Harmondsworth: Penguin Books.

Chandler, J.A. and Paul Lawless 1985. *Local Authorities and the Creation of Employment*, Aldershot: Gower.

Clark, Terry N. and Lorna C. Ferguson 1983. *City Money, Political Processes, Fiscal Strain, and Retrenchment*, New York: Columbia University Press.

Clark, Terry N., Margaret M. Burg and Martha Villegas de Landa 1984. 'Urban Political Cultures and Fiscal Austerity Strategies', Annual Meeting of the American Political Science Association, 30 August–2 September.

Clark, Terry N., Vincent Hoffmann-Martinot and Jean-Yves Nevers 1987. 'L'Innovation municipale à l'épreuve de l'austérité' budgetaire: rationalisation des services publics et contraintes politiques', report to the Housing Ministry, Oct.

Crozier, Michel 1967. 'White-Collar Unions: The Case of France', in Adolf Sturmthal (ed.), *White-Collar Trade Unions: Contemporary Developments in Industrialized Societies*, Urbana: University of Illinois, pp. 90–126.

Dion, Stephane 1986. *La Politisation des mairies*, Paris: Economica.

Downs, Anthony 1967. *Inside Bureaucracy*, Boston: Little, Brown.

Duke, Vic and Stephen Edgell 1986. 'Local Authorities' Spending Cuts and Local Political Control', *British Journal of Political Science*, 16(2): 253–268.

Eckstein, Harry 1966. *Division and Cohesion in Democracy: A Study of Norway*, Princeton University Press.

Eisinger, Peter K. 1982a. 'Black Employment in Municipal Jobs: The Impact of Black Political Power', *American Political Science Review*, 76: 380–392.

Eisinger, Peter K. 1982b. 'The Economic Conditions of Black Employment in Municipal Bureaucracies', *American Journal of Political Science*, 26(4): 754–771.

Esping-Andersen, Gösta and Walter Korpi 1987. 'From Poor Relief to Institutional Welfare States: The Development of Scandinavian Social Policy', in Robert Erikson et al. (eds), *The Scandinavian Model: Welfare States and Welfare Research*, Armonk, NY: M.E. Sharpe, pp. 39–74.

Ferman, Barbara 1985. *Governing the Ungovernable City: Political Skill, Leadership, and the Modern Mayor*, Philadelphia: Temple University Press.

Ferris, James M. 1986. 'The Decision to Contract Out', *Urban Affairs Quarterly*, 22(2): 289–311.

Fitzgerald, Michael R., William Lyons and Floydette C. Cory 1988. 'From Administration to Oversight: Privatization and its Aftermath in a Southern City', in Richard C. Hula (ed.), *Market-Based Public Policy*, London: Macmillan, pp. 69–83.

Flora, Peter (ed.) 1988. *Growth to Limits: The Western European Welfare States since World War II*, vol. 1: *Sweden, Norway, Finland, Denmark*, New York: Walter de Gruyter.

Glassberg, Andrew D. 1980. *Responses to Fiscal Crisis: Big City Government in Britain and America*, Glasgow: University of Strathclyde Centre for the Study of Public Policy, Studies in Public Policy, no. 55.

Goldsmith, Michael 1987. 'The Changing Role of Mayors and Chief Executives', ECPR Joint Sessions, Amsterdam, 10–15 April.

Goldsmith, Michael 1988. 'British Policy towards Local Government, 1979–1988', ECPR Workshop on Policy Change in Perspective, Rimini, 5–10 April.

Gramlich, Edward M. and Daniel L. Rubinfeld 1982. 'Voting on Public Spending: Differences between Public Employees, Transfer Recipients, and Private Workers', *Journal of Policy Analysis and Management*, 1(4): 516–533.

Grimshaw, William J. 1982. 'The Daley Legacy: A Declining Politics of Party, Race, and Public Unions', in Samuel K. Gove and Louis H. Masotti (eds), *After Daley: Chicago Politics in Transition*, Urbana: University of Illinois Press, pp. 57–87.

Gross, Ernest 1985. 'Labor Relations in Local and County Government', in Jack Rabin and Don Dodd (eds), *State and Local Government Administration*, New York: Marcel Dekker, pp. 409–435.

Grosskopf, Shawna 1981. 'Public Employment's Impact on the Future of Urban Economies', in Roy Bahl (ed.), *Urban Government Finance: Emerging Trends*, Urban Affairs Annual Reviews, vol. 20, Beverly Hills: Sage, pp. 39–62.

Gustafsson, Agne 1986. 'Rise and Decline of Nations: Sweden', *Scandinavian Political Studies*, 9(1): 35–50.

Gyford, John 1985. *The Politics of Local Socialism*, London: George Allen & Unwin.

Gyford, John and Mari James 1983. *National Parties and Local Politics*, London: George Allen & Unwin.

Hampton, William 1987. *Local Government and Urban Politics*, Harlow: Longman.

Hansen, Tore 1981. 'The Dynamics of Local Expenditure Growth: Local Government Finance in Sweden', in L.J. Sharpe (ed.), *The Local Fiscal Crisis in Western Europe: Myths and Realities*, London: Sage, pp. 165–193.

Hansen, Tore and Rune J. Sorensen 1988. 'The Growth of Local Government in Norway: The Sky is the Limit?' International Workshop on Local Finances in the Contemporary State: Theory and Practice, University of Oslo, 5–7 May.

Hoffmann-Martinot, Vincent 1988. 'Gestion moderniste à Nîmes: construction d'une image de ville', *Les Annales de la recherche urbaine*, 38: 95–103.

Inglehart, Ronald 1989. *Culture Change*, Princeton University Press.

Inglehart, Ronald and Jacques-René Rabier 1984. 'Du bonheur ... sentiment personnel et norme culturelle', *Futuribles*, no. 81: 3–29.

INSEE 1988a. 'Les Agents de l'etat au ler janvier 1987', *Premiers résultats*, no. 125.

INSEE 1988b. 'Les Effectifs des collectivités territoriales au ler janvier 1987', *Premiers résultats*, no. 135.

Jobert, Bruno and Michele Sellier 1977. 'Les Grandes villes: autonomie locale et innovation politique', *Revue française de science politique*, 2(2): 205–227.

Karran, Terence 1984. 'The Local Government Workforce: Public Sector Paragon or Private Sector Parasite?' *Local Government Studies*, 10(4): 39–58.

Kassalow, Everett M. 1984. 'The Future of American Unionism: A Comparative Perspective', *Annals of the American Academy of Political and Social Science*, vol. 473: 52–63.

Kiviniemi, Markku 1988. 'Local Government Reforms and Structural Changes in Public Administration: the Finish Case', in Bruno Dente and Francesco Kjellberg (eds), *The Dynamics of Institutional Change: Local Government Reorganization in Western Democracies*, London: Sage, pp. 70–88.

Ladd, Helen F. and Julie B. Wilson 1983. 'Who Supports Tax Limitations: Evidence from Massachusetts' Proposition 2 ½', *Journal of Policy Analysis and Management*, 2(2): 255–279.

Lapassat, Etienne-Jean 1977. 'Fonction publique et pouvoir d'expertise des grandes villes', in Institut Francais des Sciences Administratives, *L'Administration des grandes villes*, Paris: Editions Cujas, pp. 329–355.

Larsen, Helge O. 1987. 'Political Authority and Local Leadership', ECPR Joint Sessions, Amsterdam, 10–15 April.

Le Net, Michel and Jean Werquin 1985. *Le Volontariat: aspects sociaux, économiques et politiques en France et dans le monde*, Paris: La Documentation Française.

Levine, Charles H. 1980. 'Organizational Decline and Cutback Management', in Charles H. Levine (ed.), *Managing Fiscal Stress: The Crisis in the Public Sector*, Chatham, NJ: Chatham House, pp. 13–30.

Levine, Charles H., Irene S. Rubin and George G. Wolohojian 1981. *The Politics of Retrenchment: How Local Governments Manage Fiscal Stress*, Beverly Hills: Sage.

Lewin, David 1986. 'Public Employee Unionism and Labor Relations in the 1980s: An Analysis of Transformation', in Seymour M. Lipset (ed.), *Unions in Transition: Entering the Second Century*, San Francisco: Institute for Contemporary Studies Press, pp. 241–264.

Lewis, Gregory B. 1988. 'The Consequences of Fiscal Stress: Cutback Management and Municipal Employment', *State and Local Government Review*, 20(2): 64–71.

Lineberry, Robert L. and Ira Sharkansky 1978. *Urban Politics and Public Policy*, 3rd edn, New York: Harper and Row.

Lipset, Seymour M. 1981. *Political Man: The Social Bases of Politics*, 2nd edn, Baltimore: Johns Hopkins University Press.

Lipset, Seymour M. 1986. 'North American Labor Movements: A Comparative Perspective', in Seymour M. Lipset (ed.), *Unions in Transition: Entering the Second Century*, San Francisco: Institute for Contemporary Studies Press, pp. 421–452.

Lipset, Seymour M. and William Schneider 1983. *The Confidence Gap: Business, Labor, and Government in the Public Mind*, New York: Free Press.

Mabileau, Albert and Pierre Sadran 1982. 'Administration et politique au niveau local', in Francis de Baecque and Jean-Louis Quermonne (eds), *Administration et politique sous la Vème République*, Paris: Presses de la Fondation Nationale des Sciences Politiques.

Mayntz, Renate 1982. *Soziologie der öffentlichen Verwaltung*, Heidelberg: Müller Juristischer Verlag (Uni-Taschenbücher, 765).

Mouriaux, René 1983. *Les Syndicats dans la société française*, Paris: Presses de la Fondation Nationale des Sciences Politiques.

Mouriaux, René 1986. *Le Syndicalisme face à la crise*, Paris: Editions La Découverte.

Mouritzen, Poul Erik 1987. 'The Demanding Citizen: Driven by Policy, Self-interest or Ideology', *European Journal of Political Research*, 15: 417–435.

Mouritzen, Poul Erik 1989. 'Fiscal Policy-Making in Times of Resource Scarcity: The Danish Case', in Susan E. Clarke (ed.), *Urban Innovation and Autonomy: The Political Implications of Policy Change*, Beverly Hills: Sage.

Mouritzen, Poul Erik 1990. 'Does Politics Matter? Does Money? Observations from 15 Years in the Treadmill', prepared for the ISA World Congress, Madrid, July.

Mouritzen, Poul Erik and Kurt Houlberg Nielsen 1988. *Handbook of Comparative Urban Fiscal Data*, Odense: Danish Data Archives.

Nevers, Jean-Yves 1983. 'Du clientélisme à la technocratie: cent ans de démocratie communale dans une grande ville, Toulouse', *Revue Française de science politique*, 33(3): 428–454.

Niskanen, William A. 1971. *Bureaucracy and Representative Government*, Chicago: Aldine and Atherton.

Page, Edward 1985. 'France: From l'Etat to big government', in Richard Rose (ed.), *Public Employment in Western Nations*, Cambridge University Press, pp. 97–125.

Page, Edward and Michael J. Goldsmith 1987. *Central and Local Government Relations: A Comparative Analysis of Western European Unitary States*, London: Sage.

Parry, Richard 1985. 'Britain: Stable Aggregates, Changing Composition', in Richard Rose (ed.), *Public Employment in Western Nations*, Cambridge University Press, pp. 54–96.

Peters, B. Guy 1984. *The Politics of Bureaucracy*, 2nd edn, New York: Longman.

Peters, B. Guy 1985a. 'Sweden: The Explosion of Public Employment', in Richard Rose (ed.), *Public Employment in Western Nations*, Cambridge University Press, pp. 203–227.

Peters, B. Guy 1985b. 'The United States: Absolute Change and Relative

Stability', in Richard Rose (ed.), *Public Employment in Western Nations*, Cambridge University Press, pp. 228–261.

Pickvance, Christopher 1985a. 'Crise économique et transformation du pouvoir local en enjeu politique: Grande-Bretagne, 1979–1984', *Anthropologie et Sociétés*, 9(2): 25–54.

Pickvance, Christopher 1985b. 'Le Gouvernement local anglais en souffrance', *Les Annales de la recherche urbaine*, no. 28: 80–93.

Pickvance, Christopher 1986. 'The Crisis of Local Government in Great Britain: An Interpretation', in M. Gottdiener (ed.), *Cities in Stress: A New Look at the Urban Crisis*, Beverly Hills: Sage, pp. 247–276.

Rhodes, R.A.W. 1984. *Corporatism, Pay Negotiations and Local Government*, Essex Papers in Politics and Government, no. 19. Colchester: University of Essex .

Rhodes, R.A.W. 1986. *The National World of Local Government*, London: George Allen & Unwin.

Roniger, Luis 1981. 'Clientelism and Patron-Client Relations: A Bibliography', in S.N. Eisenstadt and René Lemarchand (eds), *Political Clientelism, Patronage and Development*, Beverly Hills: Sage, pp. 297–330.

Rose, Richard 1985. 'The Significance of Public Employment', in Richard Rose (ed.), *Public Employment in Western Nations*, Cambridge University Press, pp. 1–53.

Rubin, Irene S. 1982. *Running in the Red: The Political Dynamics of Urban Fiscal Stress*, Albany: State University of New York Press.

Sårlvik, Bo and Soren Holmberg 1985. 'Social Determinants of Party Choice in Swedish Elections, 1956–1982', IPSA Congress, Paris.

Sayre, Wallace S. and Herbert Kaufman 1965. *Governing New York City: Politics in the Metropolis*, New York: W.W. Norton.

Schneider, Mark 1988. 'The Demand for the Suburban Public Work Force: Residents, Workers, and Politicians', *Journal of Politics*, 50(1): 89–107.

Schwerin, Don S. 1980. 'The Limits of Organization as a Response to Wage–Price Problems', in Richard Rose (ed.), *Challenge to Governance: Studies in Overloaded Polities*, Beverly Hills: Sage, pp. 73–106.

Shefter, Martin 1985. *Political Crisis, Fiscal Crisis: The Collapse and Revival of New York City*, New York: Basic Books.

SOFRES 1987. *L'Etat de l'opinion. Clés pour 1987*, Paris: Seuil.

Stein, Robert M. 1984. 'Municipal Public Employment: An Examination of Intergovernmental Influences', *American Journal of Political Science*, 28(4): 636–653.

Stein, Robert M., Elizabeth G. Sinclair and Max Neiman 1986. 'Local Government and Fiscal Stress: An Exploration into Spending and Public Employment Decisions'. in M. Gottdiener (ed.), *Cities in Stress: A New Look at the Urban Crisis*, Urban Affairs Annual Reviews, vol. 30, Beverly Hills: Sage, pp. 111–144.

Stern, James L. 1984. 'A Look Ahead at Public Employee Unionism', *Annals of the American Academy of Political and Social Science*, vol. 473, May, pp. 165–176.

Stieber, Jack 1973. *Public Employee Unionism Structure, Growth, Policy*, Washington, DC: Brookings Institution.

Stoetzel, Jean 1983. *Les Valeurs du temps présent: une enquête européenne*, Paris: Presses Universitaires de France.

Thoenig, Jean-Claude 1982. 'La Politique de l'Etat à l'égard des personnels des communes (1884–1939)', *Revue Française d'administration publique*, 3: 487–517.

Thoenig, Jean-Claude and François Dupuy 1980. *Réformer ou déformer la formation permanente des administrateurs locaux*, Paris: Cujas.

Thomson, Andrew. 1982 'Local Government as an Employer', in Richard Rose and Edward Page (eds), *Fiscal Stress in Cities*, Cambridge University Press, pp. 107–136.

Tonboe, Jens Chr. 1986. 'The Labor Movement, Local Politics and Spatial Sociology: Some Recent Danish Experience', in Michael Goldsmith and Soren Villadsen (eds), *Urban Political Theory and the Management of Fiscal Stress*, Aldershot: Gower, pp. 179–205.

Troy, Leo 1986. 'The Rise and Fall of American Trade Unions: The Labor Movement from FDR to RR', in Seymour M. Lipset (ed.), *Unions in Transition: Entering the Second Century*, San Francisco: Institute for Contemporary Studies Press, pp. 75–109.

Tullock, Gordon 1970. *Private Wants, Public Means: An Economic Analysis of the Desirable Scope of Government*, New York: Basic Books.

US Department of Commerce, Bureau of the Census. 1988a. *City Employment in 1986*, GE-86-2, Government Employment, Washington, DC: US Government Printing Office.

US Department of Commerce, Bureau of the Census. 1988b. *Public Employment in 1986*, GE-86-1, Government Employment, Washington, DC: US Government Printing Office.

Walker, David 1983. *Municipal Empire: The Town Halls and Their Beneficiaries*, Hounslow: Maurice Temple Smith.

Weber, Max 1963. *Le Savant et le politique*, Paris: Plon.

Whiteley, Paul 1986. 'Predicting the Labour Vote in 1983: Social Backgrounds Versus Subjective Evaluations', *Political Studies*, 34(1): 82–98.

Wilson, Frank L. 1983. 'Les Groupes d'intérêt sous la Cinquième République: test de trois modèles théoriques de l'interaction entre groupes et gouvernement', *Revue Française de science politique*, 33(2): 220–254.

Wolman, Harold 1983. 'Understanding Local Government Responses to Fiscal Pressure: a Cross National Analysis', *Journal of Public Policy*, 3(3): 245–263.

Woody, Bette 1982. *Managing Crisis Cities: The New Black Leadership and the Politics of Resource Allocation*, Westport, Conn.: Greenwood Press.

Wynne, George G. 1983. *Learning from Abroad Cutback Management: A Trinational Perspective*, New Brunswick, NJ: Transaction Books.

Ylönen, Ari 1985. 'Societal Factors Shaping the Internal Structure of Finnish Cities', in Willem van Vliet, Elizabeth Huttman and Sylvia Fava (eds), *Housing Needs and Policy Approaches: Trends in Thirteen Countries*, Durham, NC: Duke University Press, pp. 291–304.

Ylönen, Ari 1986. 'Public/Private Partnerships: A Lesson to be Learned?' ECPR Workshop on Political Learning, Fiscal Austerity and Urban Innovation, Goteborg.

11
Professional Management and Innovation

Stephen C. Brooks

Among the actors contributing to fiscal decisions of urban areas are local government administrators. Their role in adapting to fiscal stress has been described by Baldersheim (1982) as the intelligence model. In this model, the chief administrator uses his or her knowledge and information to persuade elected decision-makers to follow the administrator's preferred course of action. The administrator hopes that the 'rational' and 'intelligent' arguments based upon extensive knowledge of budgeting and local government administration will be more persuasive than those coming from interest groups, media or the public when local politicians make fiscal decisions. The intelligence model is really put to the test during times of fiscal stress.

The intelligence model rests upon two assumptions. First, professional managers have the power and ability to largely influence or direct the making of fiscal policy. This is not an unreasonable assumption when we consider the manager's role in the development of urban policy. While democratic systems are controlled by elected officials, the day-to-day operation of most cities is in the hands of professional managers. They have a direct influence on budgeting and expenditure decisions that are essential in developing strategies to cope with fiscal stress.

In fact, before the era of 'fiscal stress' forced students of micro budgeting to reconsider budgeting practices, the administrative routines and budget recommendations of financial officers were considered key variables in predicting urban budgets (Crecine, 1969). This was confirmed in studies of other nations (see, for example, Cowart and Brofoss, 1979).

Scholarly assessment of the role of administrators in times of fiscal stress has been described as a mix of possibilities. As Wolman (1986) suggests, the relationship between fiscal stress in local government and innovation is characterized by instability. Little clarification has been given about the determinants of innovation in local government – whether they are fiscally stressed or not. One possible scenario is that fiscal retrenchment will increasingly politicize the budget process, reducing reliance on the intelligence model. As Levine et al. (1981) suggest, increased politicization reduces the ability to develop innovative responses to stress. However, this politicization is often offset in the USA by centralized power in administratively dominated city manager cities. The opposite scenario is also possible. In times of fiscal stress, political leaders may rely more upon administrators to develop strategies and policy to cope with limited resources. The significant role of the chief executive in determining innovation is well demonstrated by Bingham et al. (1981). Here, the administrator takes the lead in demonstrating how to respond to fiscal stress.

In almost all systems, administrators play a dual role in making fiscal policy. On the one hand, they often 'direct' the initial formulation and drafting of the budget. On the other hand, they usually serve as the main source of information about budgeting and spending for elected decision-makers. Because of this direct involvement in making urban policy, a knowledge of the policy perceptions and predilections of administrators will provide an understanding of their role in urban fiscal decision-making.

The second assumption in the intelligence model is that, as the name implies, this model is based upon some 'rational' process of making the choices involved in adapting to fiscal stress. This too is not an unreasonable assumption. More than any other actors in the politics of fiscal decision-making, administrators are most likely to regard the process as needing to be rational. This perception has a number of sources. First, their positions are ones of management – organizing and allocating resources. Their day-to-day activities and problems focus on how to implement policy using the money and personnel given to them. Such activity leads to developing a rationale for decision-making which, if not rational by the theoretical definition, is to some degree consistent and justifiable.

Second, many administrators have been trained as public administrators. The historic development of the discipline is based on the belief that, with better information, more sophisticated decision-making techniques and well trained decision-makers, intelligent, non-political – and therefore the 'best' – decisions can be made. For many administrators, their training has its roots in the movement toward 'scientific management' and the elimination of politics from the 'business' of running cities. A belief in professionalism is reflected in the growth of public management and public administration degree programs and, more importantly, in the assumption that urban administrators need such training to be qualified for their positions. Mohr (1969) has indicated that professionalism, among other individual characteristics, is related to organizational innovation.

These influences, combined with pressures from political leaders and the public for ways to evaluate urban administrators, often lead administrators to use 'efficiency' as the yardstick of administrative performance. By instituting new mechanical or management techniques, greater efficiencies can be found, thus increasing productivity. From this perspective, one of the main tasks of a public administrator is to have an efficient administration.

It is not surprising, then, that a common response of administrators to fiscal stress has been to resort to the use of administrative solutions to cope with the problems. By providing the 'lead' in responding to fiscal stress, administrators can continue to rely on the intelligence model for fiscal policy decisions. Using 'efficiency' to overcome limited resources is an expected response by those in management. One US study has shown that administrators assess their use of efficient management to be much higher than more objective measures substantiate (Ammons and King, 1983).

Of course, while the use (or believed use) of the intelligence model by administrators can be expected, it does not automatically happen, nor is it always a success. An administrator may not have the information he or she needs to make the best arguments, or the 'intelligent' argument may not be well received. Likewise, 'rational' administration and improvements to efficiency can be hampered by limited resources, lack of approval from political leaders or the structure of the system.

A first step in investigating the existence of the intelligence model is to ask how public administrators see their decision-making environment. With our data we cannot fully test the intelligence model in the countries studied. However, our data can give a very clear picture of how professional administrators envision their policy environments and how they believe they are coping with fiscal stress. We can develop a description of the policy maps used by administrators and can discover the role of certain administrative techniques in the development of those maps. If we assume that those who use the intelligence model are more likely to push for 'better' information, a comparison of administrators' use of sophisticated management techniques city by city can provide an initial view into the role of management in fiscal policy-making.

MEASURING MANAGEMENT SOPHISTICATION

The Fiscal Austerity and Urban Innovation (FAUI) questionnaire asked chief executive officers to rate the extent to which their local governments had instituted seven different management techniques. While the wording of the questions varied across countries, they were all designed to measure the use of these techniques. They all asked questions on a continuum to reflect not simply the use of practices but the level of sophistication in using them.

Revenue forecasting　One method of preparing for the roller-coaster changes during fiscal stress is to predict revenues. The continuum used in this question ranges local government use of forecasting techniques from using basic incremental calculations to predict next year's revenue on one end to the use of computer models on the other. (See the sample questionnaire in Appendix 11A at the end of this chapter.)

Fiscal information systems   While revenue forecasting concerns income, this question relates to local government sophistication in examining expenditure. It uses a continuum ranging from infrequent (i.e. quarterly or less) monitoring of expenditure at one end to weekly computerized reports at the other.

Performance measures   In the quest for efficiency, one technique that has been often used is developing workload and/or effectiveness measures to evaluate departmental efficiency. This question asks the chief administrative officer to rank the use of these techniques, ranging from no use of them to the extensive use of them in budgeting and management.

Accounting and financial reporting   This question taps the accounting sophistication of local government. It asks the extent to which the local officials use accepted and recommended accounting practices for government. The possible range goes from only meeting legal requirements at one extreme to receiving professional recognition for excellent accounting practices at the other.

Management rights   In both the public and private sector, cost savings can be achieved by realigning working practices of employees. This question asks about the extent to which management has control over the determination of these practices. The range is from almost all decisions being a part of employee collective bargaining and contracts, to sole management control over working practices.

Economic development impacts   As the local economy may have an influence on local government finance, this question asks about the extent to which administrative officials monitor changes in the local economy. The continuum goes from no analysis of the local economy to several ongoing analyses conducted at least annually.

Reporting to council   This question concerns the extent to which reporting on financial matters to political leadership is used by administrators. The range is from reporting to council only at budget time to reporting at each council meeting.

Index   The FAUI project asked all country teams to develop an index of management sophistication using five of the techniques described. Not all seven of the above aspects of city management are easily controlled by administrators: management rights and reporting to council can often be controlled by groups or individuals over which the professional manager has no control. The remaining five, however, are more directly controlled by local management and are therefore included. The index simply adds the level of sophistication in revenue forecasting, fiscal information systems, performance measures, accounting and financial reporting, and economic development impacts and divides by 5.

The questionnaire administration varied between the five countries reported here. For specific details see the Technical Appendix. Surveys were administered in 1985 in France, Finland and Norway. USA surveys were completed over the 1983–84 period and UK surveys were administered in 1987.

While the questionnaires tapped the same dimensions, they used various scales. Norway and Finland used a 3-point scale, France a 4-point scale and the UK and the USA a 5-point scale (see Appendix 11A). To develop comparability, all statistics presented here have been standardized to a 100-point scale.

RESULTS

Before making cross-national comparisons of these measures of management sophistication, some cautions should be noted. First, it should be remembered that these are not statements of actual practices but the perceptions of administrative officers of the extent to which these techniques are used. This difference exists whenever informants are used in research. Certainly chief administrative officers are the most knowledgeable informants one could find for information on this topic.

When considering cross-national comparisons, it is important to realize the difference in time when the surveys were administered. Those governments surveyed early may have become more sophisticated over time. This is especially true for those questions implying the use of computers. Rapid development of computer hardware and software means that, for example, computerized revenue forecasting was much more accessible to an administrator in 1987 than it was in 1983.

In addition, questions may reflect country biases. Legal requirements for management practices will vary from country to country. Some administrators may score high on some scales simply because they have an incentive to do so by the external environment or law. A good example is economic development impacts. A local government whose income is based on the success of the local economy (income or sales tax) is more likely to monitor local economic development than a local government dependent on national government revenues.

Table 11.1 presents the standardized mean of the management sophistication index for five FAUI countries. (On standardization, see the Technical Appendix.) Three countries – Finland, the UK and the USA – have similar scores of around 60. Evidently there exists a relative high degree of homogeneity in Finland and Britain, as there are smaller differences between local governments in these countries with respect to

Table 11.1   *Management sophistication index (standardized score)*

	Index	Coeff. of variation[1]	Cases (N)
Norway	29	0.24	276
Finland	61	0.14	73
UK	65	0.17	81
France	46	0.24	121
USA	61	0.21	450

[1] Original standard deviation divided by original mean.

management sophistication. France has a smaller score of 46, and Norway an even smaller one with a mean of 29. Within the last two countries there seem to be relatively big differences between localities concerning the introduction of management techniques.

The variation in scores across countries indicates that the use of these techniques is regarded as being more limited in France and Norway than in the other three countries; their administrators report that they do not rely upon these activities when planning and evaluating fiscal policy.

A better understanding of the variation can be gained from disaggregating the scores of the index. Table 11.2 presents the standardized mean score for each management sophistication item for the FAUI countries. The scores show that, as the means indicated, use of these techniques is reported to be more limited in France and Norway than in Finland, Britain or the USA.

Across the countries, however, the popularity of some techniques over others can be seen. The scores indicating highest use are for accounting and financial reporting, ranking the first or second highest in all countries. While these practices may vary from country to country, some financial accountability is necessary. In some countries this function is mandated by law.

Table 11.2   *The individual management sophistication techniques (standardized scores)*

	Norway	Finland	UK	France	USA
Revenue forecasting	43	92	59	45	61
Fiscal info systems	30	31	65	69	69
Performance measures	23	60	48	37	48
Accounting/financial reporting	36	84	83	50	67
Economic development impacts	18	NA	61	32	61
Report to council	30	38	64	NA	65

Next most popular are systems for forecasting revenues or tracking expenditures. Fiscal management decisions need to be based on information about income and expenditure. The more sophisticated these systems are, the easier it is to plan future budgets. For Finland and Norway, chief executives assess revenue forecasting techniques as most popular. For France, the UK and the USA, tracking expenditure is regarded as more popular.

The lowest scores are in the use of performance measures and the assessment of economic development impacts. The first, while touted as a good management information tool, is not easy to implement and is often a costly practice. The second, while helpful, does not provide specific information to administrators about the state of the budget. The most popular techniques reported, then, are those that provide specific information on income and expenditures.

So far, examination of management techniques has described only the reported levels of sophistication of various local governments. Of greater interest is *why* the variation exists and, more specifically, whether the variation has anything to do with fiscal stress.

Table 11.3 presents the relationship between reported use of management techniques (the overall score of Table 11.1) and the degree of fiscal slack. (For a complete description of fiscal slack, see Chapters 3 and 4.) The first two columns show the average score for each country, while the final column provides the correlation between the management sophistication index and fiscal slack. These correlations are developed from city-level data. A positive correlation means that, within each country, those local governments with greater slack also have greater perceived management sophistication. Likewise, a negative correlation means that, in those cities with greater fiscal stress, chief

Table 11.3   *Use of management techniques, fiscal slack score and sophistication/slack correlation*

	Management sophistication index	Fiscal slack[1] score	Pearson's correlation[2]
Norway	29	123	−0.07*
Finland	61	131	−0.02
UK	65	91	NA
France	46	121	NA
USA	61	95	−0.06

[1] Fiscal slack is measured in 1984 with 1978 as the base year; cf. Chapter 3 and Table 4.1.
[2] Correlation between score on management index and fiscal slack at the city level.
* Significant at the 0.05 level.

executive officers reported using more sophisticated techniques.

Table 11.3 shows only weak relationships between the management sophistication index and fiscal slack. Only one correlation is statistically significant. All three correlations are negative, suggesting that those localities with low slack (high stress) reported more sophisticated management. But the size of the correlation coefficients prevents us from drawing any definitive conclusion about the relationship.

However, looking beyond the statistical relationships, an interesting pattern can be seen in the table. As described in Chapter 4, of these five countries, local governments in the UK and the USA were the most fiscally stressed, with local governments in Finland, France and Norway in positions of fiscal slack. If Finland is viewed as a deviant case, a pattern can be seen between fiscal slack on one hand and the use of management techniques on the other. This aggregate country-level pattern supports the direction of the city-level correlations; that is, fiscal slack is negatively correlated with degree of administrative sophistication.

But why are these relations not stronger? Certainly, it may be due to the fact that no relationship exists between fiscal slack and management practices (actual or perceived). However, it may also be due to our conceptualization of fiscal slack as an 'external event' (socio-economic imbalance; cf. Chapter 3) which turned out to be largely dependent on intergovernmental revenues. It is highly unlikely that the introduction of sophisticated management techniques could reduce fiscal stress if stress is under the control of other agencies, as is the case with intergovernmental revenues. In Chapter 3 a qualitatively different conceptualization of fiscal crisis was discussed, the maladaptation concept, which focused on the extent to which the local private sector was able to support public expenditure; if fiscal policies did not reflect available resources, the result was maladaptation. It is the application of forecasting systems that is likely to affect the the degree of maladaptation, not the harshness of socio-economic imbalances created by reductions in grants or a depression in the local private sector.

Even if we were looking at management sophistication from a maladaptation perspective, there is not necessarily any link between management sophistication and the fiscal health of a city. Danish financial officers have for many years had access to sophisticated revenue forecasting systems developed over a long period by the local-government-owned computer service bureau, Kommunedata; also, several Danish municipalities have developed their own systems. It is quite clear from their experiences of the 1980s that these systems did not matter very much as soon as the local fiscal environment became turbulent. The downturn in the beginning of the 1980s, the boom of 1985 and 1986 and the next economic downturn from 1987 were all events that no sophisticated forecasting system could predict; and far less, of course, could the dramatic reductions in general grants to local governments have been predicted.

On top of that comes political manipulation with revenue forecasts. In all municipalities it is a political decision to set next year's tax base, a decision that is very often based on the necessity to reach a quick compromise rather than any kind of 'intelligence model' of fiscal policy-making. This mechanism and its long-term negative consequences is well-known to politicians as well as administrators. Nevertheless, it is used quite extensively, particularly in election years, to hide the costs of popular decisions.

It also could be argued that, instead of sophisticated management reducing fiscal stress, the causal direction goes the opposite way: fiscal stress produces more sophisticated management. Although the evidence is still rather weak, our results tend to support the hypothesis that innovation in local government increases as a product of austerity. The opposite hypothesis, that innovation follows from slack resources, could not find any support at all.

Although we cannot conclusively demonstrate the causes of perceived management sophistication, survey responses demonstrate a consistency in the belief in and reported use of management techniques by administrators. Included in the questionnaire to these officials was a list of strategies that they might use in response to fiscal stress (see Chapter 9). While the exact number varies from country to country, approximately 30 possible strategies were provided and questionnaires were designed to have comparability.

Examples of strategies included 'Across-the-board cuts', 'Increase fees and charges', 'Increase short-term borrowing', 'Contract out services', 'Impose a hiring freeze', etc. The administrator was asked to rate how important each strategy had been for his or her local authority. Again, this does not reflect the actual use of these strategies but rather the officer's perception of each strategy's importance.

Responses to these questions were then collapsed into five indexes. The revenue strategies index summed the responses to those strategies that called for increasing revenues and divided that sum by the total number of strategies providing a mean response. The expenditure strategies index used the same procedure for strategies that called for expenditure cuts. Personnel reduction strategies were those calling for eliminating personnel. Management strategies included those strategies that brought about savings through productivity increases. Finally, 'contracting out' referred to strategies where services were contracted to or shared with other government agencies or the private sector. (See Chapter 9, particularly Table 9.2.)

If the managers in our surveys believe in professional management, it is likely that those most willing to use sophisticated techniques are those

Table 11.4 *Correlation of management sophistication index and fiscal austerity strategies (Pearson's r)*

	Norway	UK	USA
Increase revenues	0.18*	0.10	0.05
Reduce expenditure	0.16*	0.16	0.11
Personnel reductions	0.14*	0.16	0.05
Management strategies	0.38**	0.18	0.34**
Contracting out	0.12	0.17	0.30**

* Significant at the 0.05 level.
** Significant at the 0.01 level.

who also believe in using management strategies to cope with fiscal stress. This consistency is demonstrated in Table 11.4 which shows the correlations between the management sophistication index and the five strategy indexes for each country. The relationships were generally weak in the UK. However, relationships do exist between management sophistication and some strategies in Norway and the USA. The strongest relationship in all three countries is between management strategies and the management sophistication index. This is not surprising. We would expect those who report using more sophisticated techniques to also report making use of a variety of strategies to cope with fiscal stress and that this use makes a difference.

CONCLUSION

The picture of administrative activity in fiscal policy-making given in this analysis shows both variety and consistency. The comparison of management sophistication suggests that urban administrators in Norway and France do not see themselves as having adopted the survey's list of management techniques to the degree of those in Britain, Finland and the USA.

However, responses within all countries show that the most used management techniques are in those areas that provide specific information on income and expenditure. The use of revenue forecasting, fiscal information systems and accounting measures rank as the most used techniques in all five countries. More esoteric management techniques like performance measures and monitoring economic development impacts are less popular. Such responses show that administrators' concerns are focused on information needed for direct fiscal decisions. If the intelligence model is used, these administrators see themselves as attempting to increase their ability to gather information for fiscal planning.

Anecdotal evidence suggests that these management techniques are reported to be used more extensively in communities in countries experiencing greater urban fiscal stress. A similar relationship may exist for individual communities. An argument can be made that those communities facing fiscal stress will institute more sophisticated monitoring of income and expenditures in an attempt to 'keep ahead' of the problems caused by fiscal stress. However, only few and weak relationships were seen between the measure of fiscal slack and the perceived use of these management techniques. This was true at the country as well as the city level of analysis.

Finally, those administrators who reported using sophisticated management techniques also were more likely to report using productivity-enhancing responses to fiscal stress. This suggests that there is a consistency in attitudes and perceptions among administrators. Those who devote time and energy to improving management techniques also see those techniques as an effective weapon against fiscal stress. It could also be argued that those administrators using these techniques as opposed to the more 'political' strategies of cutting expenditure or increasing revenues are stronger adherents to the intelligence model of fiscal policy-making, relying more on providing information and rational responses to fiscal stress.

As with most cross-national surveys of a broad topic, much of this analysis asks more questions than it solves. Certainly, more consideration needs to be given to the cultural differences among these countries as explanations for cross-country variation. Closer inspection of survey responses and case studies of administrative behavior can reinforce the speculative conclusions made here. However, it is clear that administrators play an important role in fiscal policy-making. The extent to which they use innovative practices helps dictate their success in using information as a management and political tool.

APPENDIX 11A

The following two pages show Question 9 of the Chief Administrative Officer questionnaire (Clarke, 1989: 270–1).

Q9   Where would you place your city on the following items? (Circle the number most appropriate for your city)

	1	2	3	4	5
a. Revenue Forecasting	NO FORMAL FORECAST. WE USE LAST YEAR PLUS OR MINUS AN INCREMENT.		A SEPARATE FORECAST FOR EACH REVENUE SOURCE, WITH EXPLICIT CRITERIA FOR EACH SOURCE.		MULTI-YEAR AS WELL AS ANNUAL FORECASTING, USING COMPUTER SOFTWARE
b. Fiscal Information System	DEPARTMENT EXPENDITURES CENTRALLY MONITORED QUARTERLY OR LESS.		DEPARTMENT EXPENDITURES MONITORED AT LEAST MONTHLY AND DEPARTURES FROM BUDGET QUESTIONED BY FINANCE STAFF.		COMPUTERIZED SYSTEM IS USED TO MONITOR SPENDING BY ALL DEPARTMENTS ON A WEEKLY BASIS
c. Performance Measure	NO PERFORMANCE MEASURES USED.		FAIRLY SPECIFIC WORK LOAD MEASURE (E.G., TONS OF GARBAGE COLLECTED; HOURS OF POLICE PATROL) AND SOME EFFECTIVENESS MEASURES (E.G., PERCENT OF CITIZEN REQUESTS COVERED; POLICE RESPONSE TIME).		ALL DEPARTMENTS USE WORK LOAD AND EFFECTIVENESS MEASURES ANNUALLY AND COMPUTE COSTS OF SERVICE PROVISION ON A REGULAR BASIS
d. Accounting and Financial Reporting	MEET BUT DO NOT EXCEED THE REQUIREMENTS OF STATE AND LOCAL LAW.		CHANGE ACCOUNTING PROCEDURES ONLY AT SPECIFIC RECOMMENDATION OF OUTSIDE AUDITORS.		HELD MFOA CERTIFICATE OF CONFORMANCE FOR OVER FIVE OF LAST TEN YEARS.

	1	2	3	4	5
e. Management Rights		NON-MANAGEMENT EMPLOYEES, THROUGH THEIR REPRESENTATIVES OR CONTRACTS, HAVE FORMALIZED INFLUENCE ON MOST DECISIONS CONCERNING WORK SCHEDULES, CREATING OR ELIMINATING POSITIONS AND LAYOFFS.		MANAGEMENT MUST CONSULT WITH EMPLOYEE REPRESENTATIVES IN ABOUT HALF THESE DECISIONS.	MANAGEMENT POSSESSES THE SOLE RIGHT TO MAKE THESE DECISIONS.
f. Economic Development Impacts	NO SPECIFIC ANALYSIS DONE OF IMPACTS CHANGING ECONOMIC BASE.		OCCASIONAL STUDIES CONDUCTED OF MAJOR PROJECTS (E.G., PROJECTED COSTS AND BENEFITS TO CITY OF SHOPPING CENTER).		ECONOMIC BASE SYSTEMATICALLY MONITORED FOR CHANGES IN RETAIL SALES, JOBS CREATED OR LOST. SEVERAL ANALYSES OF SPECIFIC PROJECTS CONDUCTED EACH YEAR.
g. Reporting to Council	A REPORT IS GIVEN TO THE COUNCIL ON REVENUES AND BUDGET AT BUDGET TIME, JUST ONCE A YEAR		FINANCIAL REPORTS MADE SEVERAL TIMES DURING YEAR.		FINANCIAL REPORT MADE AT VIRTUALLY EVERY COUNCIL MEETING

REFERENCES

Ammons, D. and J. King 1983. 'Productivity Improvement in Local Government: Its Place Among Competing Priorities', *Public Administration Review*, 43: 113–120.

Baldersheim, Harald 1982. 'Administrative Leadership in a Big City', in G.M. Hellstern et al. (eds), *Applied Urban Research: Proceedings of the European Meeting on Applied Urban Research*, Vol. II, Essen: Bundesforschungsanstalt für Landeskunde und Raumordnung.

Bingham, R., B. Hawkins, J. Frendreis and M. Leblanc 1981. *Professional Associations and Municipal Innovation*, Madison: University of Wisconsin Press.

Clarke, S.E. (ed.) 1989. *Urban Innovation and Autonomy*, Newbury Park, Cal.: Sage.

Cowart, A. and K. Brofoss 1979. *Decisions, Politics and Change*, Oslo: Universitetsforlaget.

Crecine, J. 1969. *Governmental Problem Solving: A Computer Simulation of Municipal Budgeting*, Chicago: Rand McNally.

Levine, C., I. Rubin and G. Wolohojian 1981. *The Politics of Retrenchment*, Beverly Hills: Sage.

Mohr, L. 1969. 'Determinants of Innovations in Organizations', *American Political Science Review*, 63: 111–126.

Wolman, H. 1986. 'Innovation in Local Government and Fiscal Austerity', *Journal of Public Policy*, 6: 161–178.

PART V

PERSPECTIVES

12

Money, Politics or Structure?

Poul Erik Mouritzen and Ari Ylönen

In this final chapter we will summarize the major findings in the volume
and relate those findings to the overall perspectives that guided the
analyses. The following four questions guided most of the analyses in
the volume.

1 What are the major determinants of fiscal slack and its opposite, fiscal stress?

Our starting point was a diagnosis of fiscal crises as emanating from
political, psychological and organizational processes in local government.
However, our analyses rested on an economic operationalization of fiscal
slack which was based on objective changes in the fiscal environments of
cities, the local tax base, grants, needs and inflation.

2 Does money matter?

It was expected that 'money matters', not only in the traditional sense,
in that the output of local political systems is a function of the resource
base, but more importantly in the sense that the whole translation
process changes character. Because the fiscal conditions differed quite
substantially between countries, this was one of the rationales for a
comparative study.

3 Does politics matter?

To what extent are public policy and political processes a function of
the distribution of power between the political parties? To answer this

question, we built on a long tradition of policy output studies; however, we extended those studies by including the potentially highly politicized issue of grant distribution, an issue where friends and foes are highly visible. Also, we extended those studies, which have almost always been of the within-country type, in attempting to estimate the same models of city fiscal policy-making processes in several countries.

4   Does structure matter?

Finally, we expected to find major differences between countries or, more precisely, between groups of countries. In this respect we distinguished three different models of local government: the Scandinavian or north European model, the southern European model, and the North American model. These models differed along a number of dimensions such as the scope of local government, the degree of consolidation, and political culture.

Clearly, the number of observations was too small to reach any definite conclusions. In the aggregated analyses we used observations from 10 countries, but in many cases information was available for only a subset of countries. What we can hope to do, therefore, is mainly to point out hypotheses that deserve further tests in future studies of local politics. Each of the four major questions is discussed below.

WHAT DETERMINES THE FISCAL WELL-BEING OF LOCAL GOVERNMENTS?

The study performed in the late 1970s by Newton and associates (Newton, 1980) showed great variation in fiscal conditions among local governments in western Europe. Looking at the fiscal conditions in the following decade, we find the same variation. Although some countries exhibited more or less the same economic conditions in the two decades (notably Italy and the UK), it is interesting to note that the fiscal environment in which local governments act can change quite dramatically. This was clearly true for Denmark.

Such changes are hard to predict. However, several of the contributions to this volume deal with the background of a local fiscal crisis, pinpointing economic, political and structural factors which, sometimes interacting with each other, define the fiscal well-being of localities.

First, it is important to note that the perception of crisis is likely to differ less among countries than are the objective economic conditions. This is due to the fact that the notion of crisis performs important political functions, as local governments often have to negotiate with central government concerning the amount of support they are to receive. Also, the notion of crisis can be used politically at the local level to minimize political losses as political leaders make unpopular decisions. Such mechanisms, in combination with psychological and organizational processes, often make localities paint a much darker picture than that actually supported by the objective fiscal conditions.

With this in mind, however, it is a major finding of the comparison of 10 countries that austerity is an objective fact that many local governments have to live with over extended periods, while other local governments are able to increase spending at the same time as they can reduce the level of taxation.

The cross-national analyses show, not surprisingly, a relatively strong relation between the general economic climate in a country and the fiscal well-being of local governments. This relation is, however, quite complex and does not hold unconditionally. Changes in the national economy work mainly via three channels as they affect localities: the local tax base, demands for public services (the so-called 'needs' component), and intergovernmental grants.

First, it is obvious that the local tax base is affected by the overall economic conditions in a country. However, only during shorter periods and only in a few countries did the tax base contribute to fiscal stress (cf. Table 4.2). Over the whole period, we found no country where the tax base contributed to stress.

Also, national economic trends are likely to have an effect on demands for services supplied by local governments. This effect can go in two directions. An economic boom will lead to lower demand for social welfare and related services. But booms can also lead to increasing demands for other services, for example day-care institutions when more women enter the labor market. The total effect on the demand side induced by changes in the general economic activity in a country is a function of the distribution of responsibilities between levels of governments and between the public and private sectors. Owing to the difficulties of measuring this effect cross-nationally, it had to be omitted from the analyses in this study. Only changes in the population were taken into consideration. Such changes were shown to be of minor importance for the fiscal well-being of local governments.

Clearly, the most important channel through which the growth of the national economy had a bearing on the fiscal well-being of local governments is the intergovernmental grants. Over a longer period, changes in grants are highly correlated with changes in the national economy as measured by changes in the gross domestic product (GDP). The relationship is, however, much stronger in times of slow economic growth or decline. This is probably a reflection of the fact that grants are more 'controllable' than many other items in the national budget, and therefore are an easy target of national government retrenchment efforts.

The relation between GDP and grants is indirect as it requires deliberate action by officials at the central level. It was expected that in periods of slow economic growth countries with a highly consolidated grant system would tend to reduce grants more than countries with a fragmented grant system because the political costs of cutting general grants are minor compared with the political costs imposed on national officials when they reduce conditional grants. This hypothesis could not be confirmed.

Building on the long tradition of cross-national output studies, it was hypothesized that governments of the right would reduce grants more than governments of the left, particularly in times of economic decline. This hypothesis gained some credibility.

The discussion above focuses on all local governments within a country. Another part of the story is the extent to which localities within a country, given the external conditions (like change in GDP and government color), are likely to suffer from fiscal stress (or to benefit from slack).

Inequality in the development of fiscal conditions is especially present in the USA, while in the Scandinavian countries localities experience more equal developments, mainly because of the equalizing effects of grant systems.

A trend that seems to hold uniformly across countries is that the bigger cities are most severely hit. This is due to a combination of grant change and changes in the local tax base. Although there was no consistent pattern across countries in the distribution of grant change, the overall combined effect goes in the direction of disfavoring the bigger municipalities.

Partisan politics was expected to affect the distribution of grants because national governments would reward their allies. This hypothesis was supported by the evidence from some countries (Sweden and France) but was contradicted by the evidence from others (Denmark, Norway and the USA).

Are the poor getting poorer and the rich richer? Although for the five countries studied there existed significant correlations between the degree of fiscal stress and city wealth, there were no uniform patterns across countries. With respect to grant change, a major component of fiscal stress, there was no uniform pattern either, although there was a slight tendency for left governments to favor the poorer cities.

Many parts of the puzzle still have to be solved. At this point we can state with some certainty what makes an individual local government vulnerable to fiscal stress:

> Large cities in a country experiencing prolonged periods of slow or no growth in the national economy in combination with large national budget deficits are most likely to be brought into a situation where current services cannot be maintained without increasing local taxes.

In periods of slow or no economic growth, the policies of central governments generally tend to reinforce rather than alleviate the effects that changes in the private sector economy have on the overall financial situation of localities. In other words, the crisis of the welfare state is exported from one level of government to the other, as predicted by structural theories of fiscal stress.

DOES MONEY MATTER? CONSEQUENCES OF A FISCAL CRISIS

In one sense, we have been able to detect policy responses that are, to a very large extent, a product of the degree of fiscal slack. Knowing the degree of fiscal slack, for instance, it is not surprising that local spending has increased very much in Finland and Norway, and decreased in Britain. Nor is it surprising that the extent to which localities in a country can increase expenditure and at the same time reduce the rate of taxation is a function of degree of slack. It follows from the way fiscal slack was defined.

In our cross-national city-level analyses (Chapter 8), we showed, in correspondence with most policy output studies, that, again, the degree of fiscal slack is the dominant determinant of changes in spending. We estimated the same models for the USA and the four Nordic countries. In all five cases the degree of fiscal slack significantly influenced cities' spending trends.

A corresponding finding was made in Chapter 5, where it was shown that local government tendency to engage in the replacement strategy increased with good fiscal conditions. The replacement strategy implies that a locality experiencing a reduction in grants from upper-level governments will replace the loss by raising local revenues.

At the aggregated (national) level of analysis, we showed that local governments tended to reduce expenditure in slow-growth periods, but to replace lost grants by local revenues in a more robust economy (Table 5.4). City-level analyses yield a similar result: the more fiscally stressed a city is, the less likely it is that local revenues will replace lost grants.

Apart from these broad trends, however, the results we have found are disappointing from the standpoint of simple explanations. Yet this in itself implies that there is considerable freedom for public officials to reshape their policies. They are not forced in any systematic way by local economic changes, government grants or other frequently discussed pressures. When it comes to the selection of strategies and policy outcomes (cf. Chapters 8 and 9), there is no consistent pattern to be found across countries suffering from different degrees of fiscal stress. The same lack of relationship was found in Chapter 10, where the degree of fiscal slack was shown to have little effect on the selection of personnel policies.

Finally, it was expected that the extent of administrative innovation would be related to the degree of slack. Such a relation could exist for two reasons. First, innovative procedures like the introduction of better forecasting techniques or fiscal information systems could trigger off warning signals at an early point, thereby allowing municipalities to take corrective actions to prevent a crisis. We now know that the basic cause of fiscal stress, independent of country, is reductions in grants; it is highly unlikely that local fiscal management systems can be of much help to a municipality when it comes to influencing the flow of money coming from central government. Second, administrative innovation could be perceived as an effect of stress; that is, a crisis would force bureaucrats to introduce fiscal management systems in the local decision-making process. In any case, there was no relation between degree of fiscal slack and administrative innovation.

Except for the basic choices concerning overall spending or the overall level of taxation, the mix of strategies used in any given situation by a local government seems rather unrelated to the availability of resources. Why is that the case? We shall touch upon two explanations.

First, there is the problem of validity. Many of our strategy indicators are based on reportings made by city officials, reportings which may often be distorted by the same type of political, psychological or organizational processes as those discussed in Chapter 3. The discussion of these processes led us to a skeptical view of the so-called subjective indicators of fiscal crises, exactly because of their expected low validity. It is therefore interesting to note the rather high degree of consistency between the reported strategies (survey-based) and the actual policy actions as they are registered by the official statistical agencies in each country (cf. Table 9.10). So, at least when it comes to the national aggregates, we have reason to believe that survey-based information does in fact correspond fairly well to reality.

The second explanation for the absence of a relation between the particular mix of strategies used and the degree of fiscal stress is that the fiscal constraints may be more dominant the higher the level of aggregation. The availability of resources may seem the determining factor when we look at overall spending and taxation. However, when it comes to the particular mix of responses such as the introduction of particular labor-saving techniques, revenue forecasting systems, borrowing or the exploitation of new local revenue sources, it is much more likely that personal or idiosyncratic factors are important. We return to this issue in the next section.

So money matters much at some level of analysis, less on other levels, when it comes to policy output.

It was also expected that political processes would be influenced by the degree of fiscal slack. The overall impression that follows from the analyses is that such relations are absent or at best very weak.

A policy map was defined in Chapter 6 as the perceptions of local political leaders of the problems and challenges faced by their municipalities. It was expected that these policy maps would reflect a higher level of crisis, the higher was the degree of fiscal stress. However, policy maps proved to be unrelated to the degree of fiscal pressure a local government had to face. The problems identified by political leaders had evidently nothing to do with fiscal slack.

Fiscal stress is a situation where unpopular decisions often cannot be avoided. Either spending has to be trimmed or taxes have to be raised. In this situation one would expect organized groups to become more active, whether they are acting to defend services or, like taxpayers' associations, fighting higher taxes. Fiscal stress does not generally lead to a more turbulent or hostile environment. Only for one country, the USA, were we able to find the expected negative relation between fiscal slack and pressure group activity as perceived by the local leaders. A reasonable explanation for this was that American cities were more severely hit by fiscal stress than their French and Norwegian counterparts. Mouritzen (1991: Chapter 7) found a result similar to the American one: the perceived level of pressure group activity was, albeit weakly, related to the degree of fiscal pressure. This referred to the 1979–82 period, when fiscal conditions in Danish local government were fairly favorable, somewhere in between those of the USA and France/Norway.

Chapter 7 focused on the preferences of mayors and mayors' perceptions of citizens' preferences. Mayors hold a central position in the policy-making process, and it was expected that they would quickly adjust their wishes for the future to the fiscal reality. Looking at aggregated data, there was a slight tendency for preferences for increased spending to be highest in the slack countries. It was expected that fiscal stress would lead to greater discrepancy between leaders' preferences and the preferences of citizens as perceived by the leaders. This hypothesis did not gain credibility.

Several of our hypotheses, often drawn from widely held views, were not supported. One explanation for the weak or non-existing effects of fiscal crisis is the fact that we argued for, and used, an objective measure of fiscal slack. For a number of reasons discussed in Chapter 3, this measure omitted some important aspects of the notion of fiscal crisis in local government. Perhaps these omitted aspects, which relate to subjective perceptions of crisis, are much more important for the perceptions of reality held by political leaders, for the shaping of their preferences and values and for their actual behavior. This line of thinking certainly needs more research.

DOES POLITICS MATTER?

A great debate took place in the 1960s and 1970s about the importance of party politics for public policy. The general impression from a myriad of analyses was summarized by Fried:

> Somehow, the nature of the socioeconomic environment seems more important than the nature of community politics in shaping community politics. The implication of these findings is that most forms of political activity are either futile or marginal, whether it be organizing to occupy office or organizing to influence those who occupy office. (Fried, 1975: 337)

We expanded the policy output perspective in this volume by focusing on grant distribution policies. In Chapter 5 we tested the partisan politics hypothesis 'that national governments reward their allies and penalize their opponents with respect to the distribution of changes in the amount of real grants'. We found some support for this hypothesis in France, Sweden and Britain, but no support at all in Norway, the USA and Denmark. Evidently there is no effect of government system here; nor does the grant system seem to have an effect, as we have two quite different grant systems in the 'non-partisan effect' category: Denmark with its highly consolidated and equalizing system, and the USA with a highly fragmented system which should, theoretically, leave more room for party-political maneuvers.

Also, we tried to estimate similar models of spending change behavior in several countries including two political variables: party of mayor, and socialist strength. These analyses, covering the four Nordic countries and the USA, in one sense confirmed the observation made by Fried 15 years ago: party politics is unimportant. In contrast, the socio-economic condition (degree of fiscal slack) is a major determinant of changes in urban expenditures. What is new, compared with most output studies performed previously by political scientists, is the importance of grants to local governments (as fiscal slack is to a major extent determined by changes in grants) as the dominant driving force of local spending. If we consider grants as a force separate from the socio-economic environment, our study is in another sense quite different from mainstream output studies, which have always stressed the *local* socio-economic climate as the major determinant of local public policy. Below we discuss these findings and others from the Danish fiscal austerity project (Mouritzen, 1991).

The output literature led to a number of methodological contributions which focused on the potential methodological weaknesses of local output research (Jacob and Lipsky, 1968; Lewis-Beck, 1977; Stonecash, 1980; Boyne, 1985). In the following pages we address some of the issues raised in this literature. We do so by suggesting four generalizations, each of which pinpoints the circumstances under which party politics is likely to influence the fiscal behavior of local political systems. The first two are substantive, in the sense that they focus on the context that is likely to produce party-political effects. The remaining two generalizations are methodological, because they address design and operational issues.

Generalization 1   Party politics will be particularly important when it comes to programs where the benefits are concentrated, which are strongly controlled by central government, and which are marked by a high degree of institutionalization.

In order to understand the context in which party-political effects on outputs will be present, we formulated in the Danish FAUI project a policy typology that was, at the outset, based on three dimensions: degree of concentration, degree of autonomy, and degree of institutionalization. We supplemented this with further characteristics of the eight services investigated, based on surveys with the political elite and the voters. With a few exceptions, there was a strong correspondence between the structural features of a program and the degree to which the elite and the electorate were divided along partisan lines with respect to their spending preferences: programs characterized by concentrated benefits, low autonomy and a high degree of institutionalization were generally found to divide the elite and the voters. On the other hand, political leaders as well as voters tended to be more in agreement on programs characterized by dispersed benefits (as is the case with pure public goods), a high degree of local autonomy and a low degree of institutionalization.

The policy typology was applied to a cross-section output study. In general, the party-political effects (socialists' share of seats, socialist majority and/or socialist dominance – more than 60 percent of the seats) were present in areas like schools but absent in areas like public roads, corresponding to the two extremes in the policy typology. Evidently the typology could explain with some success whether or not the partisan composition of the city council would make a difference; however, it was less successful in explaining the *strength* of the party-political component. We introduce a second generalization:

Generalization 2   Party politics will be less important the higher the level of aggregation.

The findings of the Danish FAUI project seem to support the expectation of Sharpe and Newton (1984: 12; cf. also Boyne, 1985: 490) that the 'tax effect constraints' will be dominant, when we look at total expenditures or at services that make up a large part of the total budget.

We did find party-political effects on services like schools and day-care institutions; but they were not, in comparison with the socio-economic effect (including the fiscal constraint), as strong as would be predicted from the policy typology. This may be due to the fact that these services make up a relatively large part of the municipal budget; in other words, the fiscal constraint increases as the scope of the service grows.

In relation to the findings of Chapter 9, we have to keep generalization 2 in mind. We did look at change in total spending – that is, we worked at a high level of aggregation – and it is therefore not surprising that party politics did not have any effect at all.

The two first generalizations illustrate the 'Policies determine politics' thesis proposed originally by Lowi (1972) (see also Peterson, 1981). Political processes are different from policy area to policy area, grounded in certain characteristics: degree of concentration of benefits, autonomy, degree of institutionalization, and share of total budget.

The two remaining generalizations focus on the research design and on the ways party-political variables are operationalized.

Generalization 3 Party-political effects will be more visible in synchronical than in diachronical designs.

In the Danish FAUI project we studied municipal allocation on the eight selected services synchronically (1982) as well as diachronically (1982–86). Generally, we found the strongest partisan effects in the 1982 cross-section study. This runs counter to the expectation of Anckar and Ståhlberg (1980: 199), who thought that party politics would particularly influence *changes* in fiscal behavior of political systems. If, however, this were so, it would be more reasonable to expect that such changes are accumulated over time, and as a consequence that they ought to be more visible in synchronically arranged designs – provided, of course, that the political party remains the same.

Within this frame of reference, it is again not very surprising that party-political effects did not show up in our diachronically arranged analysis in Chapter 9. Of course, this points toward the necessity to estimate cross-section models in a cross-national perspective, a design that perhaps raises more problems than it is able to solve because we have to adjust for a number of differences between countries, particularly the division of responsibilities between different layers of government.

Policy output studies have been particularly prone to criticism because of their high reliance on available statistical evidence. As a consequence, the operationalization of political system variables has often been rather primitive and selective (cf. for instance Jacob and Lipsky, 1968). In particular, it has been suggested that the reliance on structural rather than behavioral variables is likely to lead to a methodological bias against party-political effects: 'We think there is reason to believe that a choice of behavioral indicators which penetrate deeper into "politics" lends itself to a more precise and more dynamic description, thus emphasizing the impact of politics' (Anckar and Ståhlberg, 1980: 200).

In Chapter 7 we analyzed one such behavioral indicator, the spending preferences of mayors in three countries: Norway, France and the USA. In the first two we found significant differences along the traditional left–right dimension, but not in the USA. In the Danish FAUI study we also found extremely large differences between political leaders from left and right. We included preference measures in several analyses of spending behavior together with the traditional structural measures.

Information on the spending preferences of the Danish local political elite was collected in the fall of 1981. Two sets of questionnaire items were used. First, political leaders were asked whether the municipality ought to use less or more on each service, or whether they were satisfied with the present level of spending. Then, because this measure has been criticized for not taking into account the fiscal constraint, we used another technique, whereby each respondent was asked to allocate a reduction of the general grants (amounting to 1 percent of the local tax base) among services and taxes. (For further discussion of this, see Clark 1976.) In other words, the politician could propose cutbacks on one or more programs, propose increases in the local tax rate, or some combination of these measures. The answers were aggregated to the municipal level thus leaving us with two indicators which showed how popular a particular service was among the leading politicians. Further, we added two additional measures, which showed how popular each program was among the leaders of the dominant coalition in the municipality.

Despite very strong and significant differences in leaders' opinions between municipalities and between the political parties, these behavioral indicators did not significantly affect the behavior of local governments during the following period. Thus, a program that was highly popular among the political elite at the start of the period did not, ceteris paribus, do better than a program with little or no support.

A number of new structural variables were included in the Danish analyses: the 'length of the left party's tenure', socialist dominance (more than 60 percent of the seats), and a dummy for socialist majority in the cross-section study. (For a discussion of these indicators cf. Sharpe and Newton, 1984: 12ff.) Although some of these variables did marginally improve the explanations in statistical terms, the traditional party variable, socialists' share of the seats, generally seemed to produce the best explanation. In the diachronically arranged analyses of spending changes, we further introduced power shifts from left to right and vice versa, but without any consistent improvement in our explanations.

The following generalization, based on the Danish FAUI study, therefore runs counter to the expectation of Anckar and Ståhlberg:

Generalization 4 Structural indicators of politics can explain local fiscal behavior better than behavioral indicators.

DOES STRUCTURE MATTER?

In North America the basic structures of local government were established rather early in the nineteenth century. These structures remain the same, characterized by several types of local government, overlapping jurisdictions, extensive use of private or non-profit providers of public services and a rather visible system of taxation. When Americans talk about the 'reform' movement, they are referring not to efforts to change the basics of this system of local government, but to efforts, initiated around the turn of the century, to bring efficiency and honesty into city government.

In contrast, when northern Europeans talk about the 'reform' movement, they are thinking about the great amalgamation reforms of the 1960s and 1970s, when local governments were made bigger, tasks were transferred from the upper level to lower levels of government, and the public sector was generally 'streamlined' to consist of three (in some cases two) levels of governments: the central, regional and local levels. Great efforts were made at that time to increase the resource bases of local governments, often in the form of less visible income taxes, collected at source in tandem with state and regional taxes. Similar reforms have been suggested in many places in the USA, but they have always been blocked by strong political forces. Intellectually, the fragmented American system finds strong support in the political economy or public choice tradition (Ostrom, 1972).

In northern Europe the consolidated local governmental system is combined with strong central control and influence. The political culture, often reflected in decades of social democratic dominance, generally favors planning and co-ordination horizontally as well as vertically (from the state to local governments), and, comparatively speaking, these countries share a strong ideology favoring equality. The northern European model is similarly characterized by grant systems established to equalize fiscal conditions among localities. The North American system, in contrast, with tens of thousands of local government units, 'makes it virtually impossible for either state or federal government to exercise anything but the most general supervision over local government activity' (Newton, 1974: 59), a situation which in most cases is supported by the individualistic political culture whereby government in general is looked upon with great skepticism. 'Flat' taxes prevail over redistribution.

The southern European model of local government, in this study represented by France and Italy, is characterized by small-scale local government, many (often extremely small) localities, mayors with strong vertical ties, and generally a strong degree of central control and influence. Reform efforts, which have only recently been introduced in France, have aimed at a decentralization of functions down to intermediate tiers of government and some consolidation in territorial terms (cf. Chapter 2).

Reform movements obviously rest on the premiss that political and administrative structures make a difference. Within most countries a myriad of studies have investigated this theme, as with studies of reformed versus unreformed cities in the USA (cf. the classical study of Lineberry and Fowler, 1967) or extensive evaluations of grand reforms like the local government research projects of Sweden. In most cases these studies have proved that reforms do make a difference, although not always as marked a difference as expected.

The present study was not designed to examine effects before and after changes in political and administrative structures, but rather to make a cross-sectional comparison of differences across a few countries. Generally, we have found differences across countries but no coherent patterns along the lines suggested by our typology. Consider just one example where it is possible theoretically to argue for the existence of strong effects of government structures.

One of the main suggestions of the public choice school is that the logic of the political–bureaucratic decision-making process is biased in favor of public expenditure (for instance Kristensen, 1980): that the political gains of increasing public expenditure are greater than the political costs of raising the necessary revenues. Bias is due to an asymmetric process in which the benefits of specific public programs are of a private nature to public employees and recipients, while the costs are of a collective nature. One of the ways in which 'structure' or local governmental system was expected to matter related the degree of this bias. Generally, we expected beforehand that the more consolidated and centralized north European system would generally be biased toward higher expenditure; the North American system, in contrast, was expected to be biased toward lower taxes.

This hypothesis rests on a number of arguments.[1] The first argument focuses on the interaction between degree of consolidation and the system of taxation. The main benefits of public expenditure are private in nature. In contrast, the costs are public and the benefits of restricting public spending are a public good: the cost and benefits appear in changing tax rates which are consumed in identical amounts by the taxpayers (Kristensen, 1980). In a highly consolidated system, the relationship between the levels of taxation and spending for an individual program is obscure and uncertain. If one is dependent on municipal services, it may be dangerous to oppose tax increases because resource scarcity may threaten those services. This argument has an interesting parallel in an American study of voter response to three tax limitation proposals in Michigan. Freimann and Grasso (1982: 59) observe that the most drastic tax limitation proposal would cut property taxes most substantially. They continue: 'However, the further actions necessitated by this cut were left unspecified. Substantial spending cuts might have resulted but it was not clear in which programs.' They see the degree of uncertainty about the effects of tax limitations as a major determinant of voting behavior.

In a highly fragmented system, the awareness of which entity receives what level of funding is much greater. Higher or lower taxes for the school district have no direct impact on the amount of money going, for instance, to the municipal fire department. There is some evidence that public employees primarily support higher spending in their own unit of government (Courant et al., 1980), which may also suggest a difference within local communities in levels of support for various governmental units. In other words, when public employees in one sector look at the total governmental resources within their community, their preference may well be to keep the overall level of taxation unchanged (or even to reduce it) by supporting an increase in their own sector at the cost of reduced expenditure in other sectors. For these reasons, we expected the fiscal policy-making process to be biased toward lower taxes the more visible was the revenue structure and the more fragmented the local government system.

The second argument focuses on the degree of consolidation in combination with the grant system. We expected the fiscal policy-making process to be biased toward higher expenditure the higher the degree of territorial consolidation and the higher the degree of equalization that is built into the grant system. This is because dissatisfied citizens will have less to gain by moving to another jurisdiction.

Citizens can act in local politics through different types of action: choice (voting for political leaders), voice (the articulation of interests and demands through organized groups) and, finally, exit (migration to another jurisdiction in which the mix of services and taxes corresponds more closely to one's preferences).

In a territorially consolidated system with a high degree of equalization, there are only minor variations in tax–service ratios. (See Chapter 4 on comparative data.) In a fragmented system with a low degree of equalization, variations may be great even within the same community. In this situation, citizens who are dissatisfied with fiscal policies may gain high benefits for low costs by choosing exit as the course of action. To the extent that contextual factors like territorial fragmentation encourage a high degree of mobility and mobility represents a pattern of flight from a city with accompanying loss of tax base, exit may become a major cause of fiscal stress. To the extent that contextual factors help reduce the importance of exit as a viable option, they may act as buffers against this particular stress-creating force. The mere possibility of exit may also be significant. Political leaders are clearly aware of the 'vicious circle' created by flight, loss of tax base, declining services, increasing taxes and more flight, etc. Conservative fiscal policies are the traditional weapon to counter exit. Even if exit is only a potential response on the part of voters, the resulting fiscal policy-making process is likely to be biased in favor of lower taxes.

Third, we expected fiscal policy-making to be biased in favor of expenditure in countries with a high degree of consolidation because of the greater possibilities in these countries for strong interest groups to form around expenditure issues. A high degree of territorial and programmatic consolidation is likely to result in a union structure where relatively strong unions operate within the same territorial boundaries. Although they may be affiliated with different programs, they have a strong incentive to form coalitions that cut across services. In a functionally consolidated system service delivery occurs within local government institutions. It has been suggested that department heads and bureaucrats within departments are more likely to defend the interests of the service delivery employees in a setting where they are directly in charge of operations. In fact, one of the arguments for privatization (of the production of services) has been that the incentives of bureaucrats will swing towards efficiency and productivity as a result (Kristensen, 1983).

For the three reasons mentioned above, we expected the North American system of government, with its high degree of fragmentation, a rather visible system of taxation and a low degree of equalization, to be biased in favor of lower taxes; in other words, we thought that in a situation of stress US local political leaders would find it more favorable to cut expenditure than to increase taxes. The opposite was expected in the northern European countries: faced with the necessity to either reduce spending or increase taxes, local governments in this group would tend to increase taxes.

The test of this proposition is made difficult by the fact that the countries investigated experienced extremely different fiscal conditions. Among the 10 countries studied in this volume, we may take Denmark and the UK as examples of the western European model of local government and the USA as an example of the North American model. Only in these three countries can we find extended periods of fiscal stress: Danish local government experienced deteriorating fiscal conditions from 1982 to 1986, British localities were hit even more during the 1979–83 period, while in the same period US local governments also experienced stress. The seriousness of these fiscal problems is indicated by the fiscal slack indicator in Table 12.1.

Table 12.1 *Fiscal indicators during stress periods in Denmark,
the UK and the USA*

Denmark	1982	1983	1984	1985	1986
Fiscal slack	100	99	98	93	93
Tax effort	100	103	104	106	106
Expenditure[1]	100	96	95	100	101
Employees	100	101	100	100	101
UK	1979	1980	1981	1982	1983
Fiscal slack	100	95	88	85	89
Tax effort	100	107	120	123	117
Expenditure[1]	100	96	90	92	92
Employees	100	99	97	95	96
USA	1979	1980	1981	1982	1983
Fiscal slack	100	97	98	96	93
Tax effort	100	99	97	99	101
Expenditure[1]	100	101	100	100	99
Employees	100	99	96	95	92

[1] Expenditure is total expenditure (capital and current).
Sources: fiscal slack, Table 4.1; tax effort, Table 9.4; expenditure, Mouritzen and Nielsen (1988: 26); employees, Table 10.4.

It is true that tax rates have gone up in Denmark, as suggested, while in the USA they have generally declined, although not very much, as can be seen in the table. However, this happened in the USA during a period of general economic decline, as indicated by a real drop in GDP over the period of 1.2 percent. What is typical of the American situation is stability in tax effort and spending that is exhibited in the table.

In Denmark stress was triggered off by reductions in grants in a period of general prosperity: real growth in GDP stood at 14.3 percent during the 1982–86 period. (Cf. also Chapter 5 on replacement.) Except for a reduction in spending in 1983 (which was forced upon localities because grants were reduced after the local budget was passed), Danish municipalities have generally left expenditure untouched.

In Britain tax effort increased considerably from 1979 to 1983 but at the same time expenditure was reduced. This took place in a setting of extreme central government pressure to curb local expenditure in a period of a stable GDP.[2]

One may also look at changes in the municipal workforce. Here we find trends that are perhaps more in line with the kind of mechanisms we expected to find in these types of countries. Danish municipalities are very reluctant to fire people. Even during the forced reduction in spending in 1983, municipal workers were protected. In absolute terms, the British workforce did decline, but at a slower rate than expenditure. In contrast, the municipal workers in the USA are evidently less protected, as we find a major reduction in the workforce even in a period where municipal spending was at a standstill.

Obviously, our number of cases is too small and there are too many extraneous factors for us to reach any firm conclusions about the effect of government structure on the behavior of local governments. This analysis only partially confirms the thesis that 'structure does matter'. What is evident from the figures, however, is perhaps a more general pattern: no matter how a local government system is organized, when it comes to the basic choices of spending and taxation, the most important factor is 'money' – or more specifically – the national economic forces of the country.

FUTURE RESEARCH

This volume has focused on the causes and consequences of urban fiscal stress in a comparative perspective. We have analysed various phases of the policy cycle: how and why the socio-economic environment of municipalities changes; how these changes affect the environment as it is perceived by local political leaders; how political leaders respond to environmental changes in the form of various packages of fiscal policies. Underlying these analyses was the assumption that basic political and administrative structures would affect the whole translation process, in other words, the interplay between the various actors in the local political system. In one respect, we have merely been able to scratch the surface of the policy cycle and the way it is formed by the structural context in a country: most of the analyses rested on aggregated data, and in only a few cases were we able to study aspects of the policy cycle at the city level.

An obvious next step is to refine the comparative analysis with information about many cities in several countries. This will ideally require socio-economic and fiscal data that will allow us to monitor how individual cities act over a period of years, covering at least three to five years between two local elections. If these could be combined with information obtained via surveys with citizens, leaders and perhaps chief administrators, we would have an excellent opportunity to test propositions about the effect of the structural and socio-economic context on local politics.

NOTES

1   Part of the following line of argument was presented in Mouritzen and Narver (1986).
2   For the changes in GDP, cf. Mouritzen and Nielsen (1988: 47).

REFERENCES

Anckar, Dag and Krister Ståhlberg 1980. 'Assessing the Impact of Politics: A Typology and Beyond', *Scandinavian Political Studies*, 3: 191–208.

Boyne, George A. 1985. 'Review Article: Theory, Methodology and Results in Political Science: The Case of Output Studies', *British Journal of Political Science*, 15: 473–515.

Clark, Terry N. (ed.) 1976. *Citizens' Preferences and Urban Public Policy*, Beverly Hills: Sage.

Courant, Paul N., Edward M. Gramlich and Daniel L. Rubinfeld 1980. 'Why Voters Support Tax Limitation Amendments: The Michigan Case', *National Tax Journal*, 33: 1–20.

Freimann, Marc P. and Patrick G. Grasso 1982. 'Budget Impact and Voters' Response to Tax Limitation Referenda', *Public Finance Quarterly*, 10: 49–66.

Fried, R. 1975. 'Comparative Urban Policy and Performance', in F.L. Greenstein and N. Polsby (eds), *The Handbook of Political Science*, vol. 6, Reading, Mass: Addison-Wesley, pp. 305–379.

Jacob, H. and Michael Lipsky 1968. 'Outputs, Structure and Power: An Assessment of Changes in the Study of State and Local Politics', *Journal of Politics*, 30: 510–538.

Kristensen, Ole P. 1980. 'The Logic of Political Bureaucratic Decision Making as a Cause of Governmental Growth', *European Journal of Political Research*, 8: 249–264.

Kristensen, Ole P. 1983. 'Markedet som alternativt styringsmiddel i den offentlige sektor', in Bøje Larsen (ed.), *Nye styremåder i den offentlige sektor*, Copenhagen: Juristog Økonomforbundets Forlag, pp. 67–92.

Lewis-Beck, Michael S. 1977. 'The Relative Importance of Socioeconomic and Political Variables for Public Policy', *American Political Science Review*, 71: 559–566.

Lineberry, Robert L. and Edmund Fowler 1967. 'Reformism and Public Policies in American Cities', *American Political Science Review*, 61: 701–716.

Lowi, Theodore 1972. 'For Systems of Policy, Politics and Choice', *Public Administration Review*, July/August: 298–309.

Mouritzen, Poul Erik 1991. *Den politiske cyklus: En undersøgelse af vælgere, politikere og bureaukrater i kommunalpolitik under ressourceknaphed*, Aarhus: Forlaget Politica.

Mouritzen, Poul Erik and Betty Jane Narver 1986. 'Fiscal Stress in Local Government. Responses in Denmark and the United States', Terry N. Clark (ed.), *Research in Urban Policy*, vol. 2, London: JAI Press, pp. 197–219.

Mouritzen, Poul Erik and Kurt Houlberg Nielsen 1988. *Handbook of Comparative Urban Fiscal Data*, Odense: Danish Data Archives, University of Odense.

Newton, Kenneth 1974. 'Community Decision Makers and Community Decision-Making in England and the United States', in Terry N. Clark (ed.), *Comparative Community Politics*, New York: John Wiley, pp. 55–86.

Newton, Kenneth 1980. *Balancing the Books: Financial Problems of Local Government in West Europe*, London: Sage.

Ostrom, Elinor 1972. 'Metropolitan Reform: Propositions Derived from Two Traditions', *Social Science Quarterly*: 53, 474–493.

Peterson, Paul E. 1981. *City Limits*, Chicago: University of Chicago Press.
Sharpe L.J. and K. Newton 1984. *Does Politics Matter? The Determinants of Public Policy*, Oxford: Clarendon Press.
Stonecash, Jeff 1980. 'Politics, Wealth and Public Policy: The Significance of Political Systems', in Thomas R. Dye and Virgina Gray (eds), *The Determinants of Public Policy*, Lexington, Mass.: Lexington Books.

Technical Appendix

THE DATA

Data referring to two different levels of analysis have been used in this book. Some analyses are based on aggregated data, which describe the fiscal context or the behavior of all municipalities in a country. Others are based on city-level data, or information on the fiscal context or the behavior of individual cities in a country. Independent of the level of analysis, two different types of data have been used: survey data, and fiscal and socio-economic data. Although great efforts were made to make the data sets complete and comparable for each of the 10 countries, it was not possible to obtain 100 percent coverage. In the Table A1 the coverage for each country is indicated.

The data from the the UK cover only England and Wales. As it was not

Table A.1   *Data coverage in the ten countries*

	Survey data		Fiscal/socio-economic data	
	Aggregated to national level	City level	Aggregated to national level	City level
Denmark	X	X	X	X
Norway	X	X	X	X
Sweden	X	X	X	X
Finland	X	X	X	X
UK	X	X	X	
Germany			X	
France	X	X	X	X
Italy			X	
Canada	X		X	
USA	X	X	X	X

possible to obtain municipal time-series data for Germany, it was decided to include all local government data in the municipal section; so when interpreting the German data it is important to bear in mind that the municipal time series in fact covers all local governments. The German time-series data have been collected by Kurt Houlberg Nielsen using the IMF and OECD as sources. The Canadian survey data covers only cities in the four western provinces of the country.

For six countries it was possible to work with both types of data at the aggregated as well as the city level of analysis: Denmark, Norway, Sweden, Finland, France and the USA. In the UK it was not possible to collect the necessary updated city-level data. Although a survey had been conducted in West Germany, the data collected were not available for the present study. In Italy the survey was not available at the time of the preparation of the manuscript.

The collection and preparation of the aggregate data took place at the University of Odense, Denmark. The data were submitted by participants from each of the countries. The aggregate data were published separately in the summer of 1988 in P.E. Mouritzen and K.H. Nielsen, *Handbook of Comparative Urban Fiscal Data*, Odense: Danish Data Archives, University of Odense.

It is not possible in this volume to present all the data sources, definitions, etc. In the *Handbook* a section contains notes for each country, while another section contains the exact definitions on which the collection of data in each country was based. Readers are generally referred to the *Handbook* for more specific information. Whenever we have used data drawn from the *Handbook* or performed analyses based on the data in the *Handbook* this is indicated by a source note. Tables and figures compiled directly from survey data collected by the individual national FAUI teams give no specific source.

The *Handbook* and the corresponding data diskettes (LOTUS 123 files) can be obtained by writing to

Danish Data Archives
Odense University, Campusvej 55
5230 Odense M, Denmark

The *Handbook* is priced at DKr 200 (approximately US$35) including postage and handling. The diskettes are sold for an additional DKr 200.

THE PROBLEM OF STANDARDIZATION

Although each national team at the start had a standardized questionnaire, the subsequent adaption to the special national setting often resulted in minor changes which make direct comparisons difficult.

A typical problem is that the scales used may differ across countries.

In one country spending preferences could be indicated on a 3-point scale (spend more, same, less), in another country on a 5-point scale (much more, more, same, less, much less). In several contributions to this volume a scale ranging from 0 to 100 has been used to compare results, mainly means, across countries.

The general formula for the conversion is (from a memo prepared by Richard Balme for the project participants)

$$X = 100 \times (x_i - \text{min}) / (\text{max} - \text{min})$$

where

X	=	value on the common scale (0 to 100)
x_i	=	value on the actual scale used in the survey
min	=	minimum value on actual scale
max	=	maximum value on actual scale

We illustrate the use of the formula by four examples. As can been seen in Table A2, the actual minimum is always converted to a value of 0, the actual maximum to a value of 100 and the value between actual minimum and maximum to 50. The conversion formula gives a pragmatic solution to an otherwise unsolvable problem. However, the formula specified does not allow us to compare variances across countries, which are likely to vary according to the number of response categories used.

Owing to differences in the rate of inflation, all aggregated time series have been deflated. In other words, all the indices referred to in the book show real growth. For most countries we used the government final consumption deflator. The special municipal deflator was used for a few countries where available.

In almost all instances, aggregated data have been taken directly from the *Handbook of Comparative Urban Fiscal Data*. Besides the above

Table A.2   *Illustration of standardization formula*

	Country 1	Country 2	Country 3	Country 4
Actual scale				
Min.	1	0	1	3
Max.	3	3	5	15
Average[1]	2	1.5	3	9
New scale				
Min.	0	0	0	0
Max.	100	100	100	100
Average	50	50	50	50

[1] Hypothetical average for the country.

mentioned deflation procedure, the indices in the *Handbook* also reflect
changes in population; that is, in all cases they have been controlled for
population change. Thus, when an index shows change over a period for
some fiscal variable, it is always change per capita in real terms.

THE INDIVIDUAL COUNTRIES

A more detailed discussion is given below for each country. For three
countries, Germany, Italy and Canada, we only used data published in
the *Handbook*. These countries are not covered below.

Denmark

The Danish FAUI project was initiated in the fall of 1980. In a first round
of data collection, citizens and local political leaders were surveyed in
November 1981. As a consequence of the start-up of the international
project in 1982, a follow-up survey of finance directors was made in the
spring of 1983.

The citizen survey was carried out by the Danish Gallup Institute and
covered a representative sample of Danish voters ($N = 1020$). The political
elite survey included the mayor, members of the finance committee
and chairmen of the standing committee in a representative sample of
40 municipalities ($N = 249$). One of the major differences between
the Danish questionnaires and the questionnaires used in most other
countries is that political leaders were not asked about their evaluation
of citizens and group preferences for spending. The data used on citizens'
spending preferences (Chapter 7) stem directly from the citizens' survey;
however, the data on pressure group activities (Chapter 6) are based on
the responses of political leaders.

As a consequence of differences in timing, it was decided to use two
time periods for the Danish city-level analyses. In those chapters where
the responses of political/administrative leaders are studied, the fiscal
measures (such as the fiscal slack indicator) refer to the period leading
up to the year of the finance director survey, in most cases 1979–83.
Except for the last year, this period was marked by fiscal slack. For
several reasons, the period 1982–86 is more interesting. In the analyses of
fiscal slack and grant changes as well as in the investigations of spending
changes, the period under investigation in Denmark is 1982–86. This
period was marked by fiscal stress.

In the city-level analyses it is important to notice that, whenever grants
are included (as a variable or in the calculation of other variables), only
general grants are counted. This is because all conditional grants are of
the matching type, a major part of which is pensions for elderly, widows
and the handicapped. On these programs municipal costs are reimbursed
by 100 percent.

The aggregated fiscal data stem for the most part from Danmarks Statistik, Statistiske Efterretninger. Aggregate data do not include the cities of Copenhagen and Frederiksberg.

The study was supported by the Danish Social Science Research Council, SSF grant no. 14–4318.

Norway

The FAUI surveys conducted in Norway cover mayors (spring 1985), chief administrative officers (fall 1985) and committee chairmen (spring 1986). The questionnaires for mayors and chief administrative officers were administered to all municipalities ($N = 454$). The response rates were 84 and 82 percent respectively. The survey of committee chairmen was conducted in 135 municipalities selected to be representative of Norwegian municipalities in general. The chairmen of the five most important committees received questionnaires that were modelled on the councillors' questionnaires of the American survey. (The committee chairmen are also members of the local councils.) There were 465 questionnaires returned, corresponding to a response rate of 67 percent. All surveys were mailed.

Statistical information on budgets, demographic change, occupational structure, etc., comes from data banks established by the Norwegian Social Science Data Services.

The surveys were financed by the Norwegian Research Council for Applied Social Science.

Sweden

The Swedish FAUI project is based mainly on three different sets of data. One is fiscal data for all (284) cities and municipalities. The data on expenditure, fees, grants and taxes used here are from the final accounts. The second set is aggregated data, based mainly on the same kind of statistics. Finally, survey data are used in the analyses. The questionnaire was mailed to the chief economic administrators in all municipalities in December 1986 and covered most aspects of the budgetary process. The response rate was 80 percent.

Finland

The FAUI project in Finland was started in 1985, financed by the University of Tampere and the Research Council for the Social Sciences of the Academy of Finland.

Identical questionnaires were mailed to three categories of respondents: city managers, finance managers, and the chairmen of city boards. The major part of the questionnaire was adopted from the standard survey method of the international study. The survey included all 84 municipalities classified as cities. The response rates were 90.5 percent (city managers), 95.2 percent (finance managers), 82.1 percent (chairmen of the board), 79.8 percent (vice chairmen of the board) and 75.8 percent (second vice chairmen of the board). Socio-economic data covering the years 1981 and 1985 were also collected and merged with the survey data.

In international comparisons, it is necessary to notice that the city manager (also called the mayor), who is commonly regarded as responsible for his city, is a permanent local civil servant. The city manager is selected from among the candidates of the different political parties by vote in the city council, but once selected he or she is a permanent civil servant.

UK

The FAUI survey in Britain was conducted of chief administrative officers only. In late spring 1987 chief executives from 116 British local authorities were contacted by letter, asking them to participate. In June and July the surveys were mailed to local authorities selected from Greater London authorities, English metropolitan districts, English non-metropolitan districts and Welsh districts. English non-metropolitan county and Welsh county authorities were not included because of the focus on urban fiscal austerity. For a similar reason, the English non-metropolitan districts and Welsh districts that were chosen were those with populations of 130,000 or above. Scottish authorities were excluded because of the unique nature of those authorities.

There were 81 questionnaires returned, corresponding to a response rate of 69.8 percent. The distribution by type of authority of those returned is very similar to the distribution of those sent. Many questionnaires were completed by the chief executive or someone on his or her staff. Those not completed in the executive's office were most often completed by the treasurer's office or a policy research unit of the local authority.

France

The French FAUI project was launched in 1983 and was rapidly supported by various ministries and institutions including the Centre National de la Recherche Scientifique, the housing department, the Association des Maires de Grandes Villes de France and the Centre de Formation des Personnels Communaux.

In 1985 two groups of local officials were surveyed in all cities with over 20,000 inhabitants ($N = 381$): mayors and chief administrators. The general pattern of the two questionnaires was very similar to the international one. Some specific themes or topics were added, especially relating to the decentralization reform or the center–periphery relations, and a list of 38 strategies was proposed, including some omitted in other countries such as the 'municipalization' or 'publicization' of services formerly delivered by private actors.

Objective data such as personnel and budgetary information were transmitted in a standardized presentation by central institutions including the finance department, the Caisse des Dépôts et Consignations and the Institut National de la Statistique et des Etudes Economiques.

The United States
The American FAUI project was conducted in 1983 and 1984. Teams of researchers throughout the USA assumed responsibility for mailing questionnaires to urban officials in their states and regions. Cities of 25,000 and larger were surveyed. Separate questionnaires were mailed to three groups of policy-makers: mayors, council members and chief administrative officers.

Major data sets were merged into the FAUI survey files. The most commonly used for fiscal information was the annual survey of municipal finances conducted by the Bureau of the Census, Government Division. This survey includes all cities of 25,000 inhabitants and over. The County and City Data Book file was also used.

The USA underwent a major recession in 1981 and 1982 with large cities reporting double-digit unemployment rates. September 1982 is usually considered the low point. For this reason, 1980 and 1984 are compared. Since the FAUI survey was conducted in 1982–83, the survey data are relevant for comparison of changes in fiscal policy outcomes. Efforts initially were made to study 1980–82 and 1982–85 on the grounds that pre-recessionary periods might differ substantially from those following the recession. Preliminary examinations revealed that a two-year period is insufficient to identify significant policy changes. For this reason, only the 1980–84 period is included.

Index

Compiled by Jackie McDermott